D. Bartz (ed.)

Visualization in
Scientific Computing '98

Proceedings of the Eurographics Workshop
in Blaubeuren, Germany,
April 20–22, 1998

Eurographics

SpringerWienNewYork

Dirk Bartz
WSI/GRIS, University of Tübingen,
Federal Republic of Germany

© 1998 Springer-Verlag/Wien

Typesetting: Camera ready by authors
Printing: Druckerei Novographic, A-1238 Wien
Binding: Fa. Papyrus, A-1100 Wien

Graphic design: Ecke Bonk

Printed on acid-free and chlorine-free bleached paper

SPIN: 10684783

With 82 partly coloured Figures

ISBN-13: 978-3-211-83209-7   e-ISBN-13: 978-3-7091-7517-0
DOI: 10.1007/978-3-7091-7517-0

Preface

Starting in the late eighties, visualization became a well established research field within the scientific community. Methods of computer graphics were applied to data generated by all kinds of scientific applications. This data can be either measured (e.g. medical scanners, meteorology) or is generated by simulations, commonly computed on super computers.

Numerous problems emerge from all parts of the visualization process. Mostly, the methods deal with better handling of huge data, or improved presentation of the data.

In 1989, the Eurographics Working Group on Visualization in Scientific Computing initiated an annual workshop on this topic. This year (1998), the ninth workshop took place at Blaubeuren, Germany. Organized by the Computer Graphics Lab of the University of Tübingen, twenty presentations discussed a variety of problems in visualization in scientific computing.

Twelve of these presentations were selected for this volume. They are roughly structured into four topics: adaptive meshing and multi-resolution, feature extraction, flow visualization, and visualization quality.

Adaptive and multi-resolution methods. Scientific Data is computed on different kinds of grids. However, the grids of modeled or measured data are frequently not adapted to the features which are of particular interest. Furthermore, a computed visualization object frequently needs to be reduced to enable its visualization. This can be done by simplification of the visualized surfaces (Frank et al.) or by compression of the information of multi-resolution structures (Klein et al.). In order to improve the quality of isosurfaces, a grid can be adaptively refined using a trilinear reconstruction filter (Allamandri et al.), or distance fields can be applied to interpolate slices between two different contours (Schilling et al.).

The visualization of large vector fields is usually limited by the amount of data. Therefore, it is important to incorporate *feature extraction* in the visualization process. The contributions in this book focus on the examination of the accuracy of algorithmic extracted features (Reinders et al.) and on enhancing the feature of dynamical systems (Löffelmann et al.).

In *flow visualization*, particle tracing is one important visualization method. Starting from an initial position, a particle is tracked on its path through a flow field by integrating the velocity of the particle to its new position. Problems which evolve with extending this method to curvilinear grids are addressed in this book (Sadarjoen et al.). Furthermore, a method for tracing particles on sparse grids is presented (Teitzel et al.). A simulation of material transport was used by Becker et al. for visualization of time-dependent velocity fields.

The topic of the last section is *visualization quality*. Besides research on algorithms and methods for the visualization of scientific data, research is necessary how to evaluate the accuracy and effectiveness of the methods. The first paper discusses the benefits and problems of the use of stereoscopic volume rendering in a planning system for conformal radiotherapy (Hubbold et al.). Thereafter, two approaches which debate the visualization quality of visualization systems (Haase) and the effectiveness of the visualization of atomic collision cascasdes (Šroubek et al.) are presented.

Finally, I would like to thank everybody who was involved in the production of this book. First of all, the authors for their contribution; next, the members of the program committee, the reviewers, and my collegues who helped me organize this workshop. Last but not least, I thank the staff of Springer-Verlag, Wien for producing this book.

May 1998   Dirk Bartz
                                                                  Tübingen, Germany

Contents

Part I
Adaptive and Multi-resolution Methods

Data–Dependent Surface Simplification

Karin Frank, Ulrich Lang

HLRS
Computing Center University Stuttgart
Allmandring 30a, D-70550 Stuttgart, Germany

Abstract In Scientific Visualization, surfaces have often attached data, e. g. cutting surfaces or isosurfaces in numerical simulations with multiple data components. These surfaces can be e. g. the output of a marching cubes algorithm which produces a large number of very small triangles. Existing triangle decimation algorithms use purely geometric criteria to simplify oversampled surfaces. This approach can lead to coarse representations of the surface in areas with high data gradients, thus loosing important information.

In this paper, a data-dependent reduction algorithm for arbitrary triangulated surfaces is presented using besides geometric criteria a gradient approximation of the data to define the order of geometric elements to be removed. Examples show that the algorithm works sufficiently fast to be used interactively in a VR environment and allows relatively high reduction rates keeping a good quality representation of the surface.

1   Introduction

In Scientific Visualization, a common approach consists in extracting isosurfaces and cutting surfaces from a data set in order to detect essential features of the data. Especially when visualizing numerical data from high performance computing (HPC) simulations, there are often multiple data components which can be mapped onto the surfaces, e. g. by a color map. The resulting colored surfaces can be still very large. Moreover, they are often produced by a marching cubes type algorithm leading to a large number of very small triangles. Thus, the simplification of those surfaces is an important task to make interactive visualization of scientific data practicable.

Existing simplification algorithms employ geometric criteria to determine the order of the reduction process. Their goal is to obtain a coarse representation of the surface in areas of low curvature, and a finer representation in more wrinkled parts of the surface. But important features in the data components may be situated in areas of rather flat shape, e. g. turbulences in flow fields mapped on a cutting plane. This shows that in the case of data attached to the surface purely geometric criteria are not sufficient to get an appropriate representation containing less triangles without loosing essential information.

We developed a surface simplification algorithm involving data-dependent criteria, namely, a gradient approximation of the data attached to the surface.

As surfaces may be of a simple shape (think of a cutting plane or sphere) geometry related criteria may seem to the user less dominant than data-dependent criteria. Therefore, our priority criterion is a linear combination of a curvature and a gradient approximation, weighted by a user-defined parameter. The gradient approximation scheme allows both scalar and vector data to be taken into account.

We decided to stay with topology preserving methods, although the algorithm can be easily adopted to allow topology changes. But, topology changes seem to be more important in the visualization of geometric models, where parts of the scene can be at such a distance from the observer that holes or cracks get smaller than pixel size. In scientific visualization, this seems to be a less urgent issue.

Obviously, the more criteria are used to control the simplification process, the less reduction can be achieved. However, the algorithm is still able to reduce large surfaces by a factor up to 10 without blurring important features in the data. Since it has a time complexity of $O(N \log N)$, with N being the number of original vertices, it works sufficiently fast to be used interactively even in a virtual reality environment.

1.1 Related work

We do not pretend to give a complete survey on surface simplification algorithms. In the last years, this topic has been developing rather fast. For a historical overview and a vast bibliography, see the paper of A. Guéziec [10].

Our algorithm belongs to the class of geometry removal methods. Direct inspirations came mostly from papers by H. Hoppe et al. [13], [15], W. Schroeder et al. [17], B. Hamann [12], A. Guéziec [10], R. Klein et al. [14], and M. Garland and P. Heckbert [8]. In these articles, different removal strategies (vertex removal, edge collapsing, triangle removal) are proposed involving various kinds of geometric priority criteria.

Data dependencies are common in the well-developed field of data dependent triangulation (see e. g. [7],[2]). There the data is usually given as function values in vertices on the 2D plane, thus creating a height field. Unfortunately, the heuristics given in these papers are not applicable directly.

In volume data visualization, there are techniques to compress data given on a three-dimensional grid by representing areas where the data are not changing essentially in a more efficient way, e. g. using a finite element approximation [9] or other basis representations (see e. g. [22]). Since we do not have access to the data on the original grid, but only to a subset of the data defined on a 2-manifold in $\mathbb{R}^3$, these methods are not applicable, either.

2 The algorithm

We use a typical geometry removal approach, deleting one geometry element at a time. The general outline of the algorithm is as follows:

ALGORITHM:
    Preprocessing operations (retrieving connectivity information)
    Construct priority queue
    `while` (priority queue is not empty)
    { Take next geometry element
       Perform topology-preserving tests
       `if` (element can be removed)
       { Collapse geometry element
          Update data structures and priority queue
       }
    }

In the following, the main steps of the algorithm will be explained in more detail.

2.1 Geometry removal

There are basically three possibilities to remove geometry elements:

- Vertex removal
- Edge collapsing
- Triangle removal

When removing a single vertex, one has to re-triangulate the resulting hole in the surface. This operation can be avoided by using the edge collapsing approach instead. Here, the two points adjacent to the edge are merged into one (see Fig.1), and the surrounding triangles merely change their shape. The simplest and fastest way to compute the new point the edge collapses in is to take the midpoint of the edge. But for convexely shaped surfaces, this leads to flattening of the surface which is sometimes not desirable. Therefore, A. Guéziec [10] developed a method of calculating the new point according to a volume preserving strategy. However, for flat surfaces like cutting planes this additional effort is unnecessary.

If the preferred generic operation is triangle removal, again the hole in the surface resulting from removing all adjacent faces must be handled. This can be done in two ways. Firstly, it can be re-triangulated, which results in comparatively large, possibly flattened spots. On the other hand, a midpoint can be chosen according to some rule, avoiding re-triangulation. For instance, B. Hamann [12] approximates the surface locally by a bi-quadratic function and computes the midpoint as the value of the approximating function in the origin of the local coordinate system. Besides being numerically sensitive, this procedure is quite expensive. But on well-shaped grids which do not contain very small triangles it leads to surfaces of a high quality.

We implemented and compared all three approaches with respect to time and memory efficiency. As a result, we prefer the vertex removal and edge collapsing strategies to triangle removal because they allow a finer tuning of the iterative process, causing less deformation of the surface in each iteration step. Hence in the present paper, we decided not to report on the triangle removal algorithm.

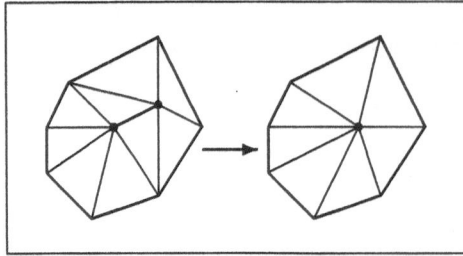 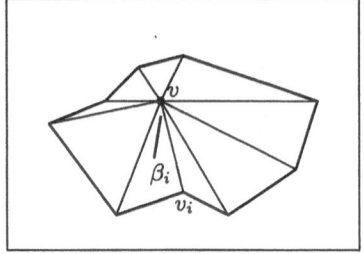

Figure1. The edge collapsing operation (left). Total angle $\Theta(v) = \sum_{i=1}^{n} \beta_i$ (right).

2.2 The data-dependent priority criterion

The heart of any geometry removal algorithm is a priority queue of geometry elements (vertices, edges, or triangles) providing the order in which they have to be removed. In existing surface simplification algorithms, the priority criterion is purely geometrical, e. g. based on a curvature estimation using an approximating function [12], [20], edge length [10], triangle area, distance to an average plane [17], or some energy function [15].

These geometric criteria are sufficient as long as no data is attached to the surface. But e. g. in the case of cutting surfaces through numerical data, which often are of a rather simple geometry, a purely geometric approach does not make sense. Therefore, we combine geometric criteria with data-dependent criteria approximating the gradient of the data.

For the sake of efficiency, we use a comparatively simple priority criterion. We combine a curvature criterion with a gradient approximation, thus assigning each vertex v a weight

$$\text{weight}(v) \;=\; \alpha * \text{curv}(v) + (1-\alpha) * \text{grad}(v)\,, \tag{1}$$

where $\alpha \in [0, 1]$ is a user-defined scalar parameter.

If the generic operation in the simplification process is edge collapsing or triangle removal instead of vertex removal, we add up the weights of all vertices belonging to the edge (or the triangle, respectively) in order to get a weight for the geometry element to be processed.

Gradient estimation Usually, the gradient of a function $f : \mathbb{R}^3 \to \mathbb{R}$ is defined as

$$\text{grad} f \;=\; \left(\frac{\partial f}{\partial x}, \frac{\partial f}{\partial y}, \frac{\partial f}{\partial z} \right)\,.$$

Since our data function is defined on an 2-manifold in $\mathbb{R}^3$ rather than on $\mathbb{R}^3$, and is not differentiable everywhere, this gradient of the data function is not defined. Hence, we have to look for other heuristics to get a measure of the rapidity of changes in the data values in the neighborhood of a vertex.

Various gradient approximation schemes for scattered data given in points on a 2D plane are described in [19]. A different approach can be imagined using again a biquadratic approximation, this time of the 3D manifold in 4D space built by the surface-attached data. Such an approximation is described in [11]. Besides being expensive, the least-square optimization used to find the best fitting biquadratic function is numerically sensitive to very small triangles which are often produced by the marching cubes algorithm. Moreover, this approximation can be used for scalar data only.

Let f_i be the data values attached to the vertices v_i, $i = 1, .., k$, adjacent to $v \in T$, and f be the data value attached to v itself. We calculate

$$\text{grad}(v) = \max_{i=1,\ldots,n} \frac{\|f_i - f\|_1}{\|x_i - x\|_1}. \tag{2}$$

Here, x_i and x denote the three-dimensional coordinates of v_i, v, respectively, and $\|\cdot\|_1$ is the usual sum vector norm assigning each vector $y = (y_1, \ldots, y_k) \in \mathbb{R}^k$ a non-negative number

$$\|y\|_1 = \sum_{i=1}^{k} |y_i|.$$

Instead of the $\|\cdot\|_1$-norm any vector norm could be used. If the Euclidean norm is used to compute the distance between vertices, this heuristic can be interpreted as an estimate for the maximal value of discretized directional derivatives

$$\frac{\partial f(v)}{\partial(x_i - x)} = (\text{grad}f(v), x_i - x),$$

taken over all directions $x_i - x$. We preferred to avoid the rather expensive square root calculation, having in mind that in $\mathbb{R}^3$ all norms are equivalent. Note that this gradient heuristic is independent of whether the data attached to the surface is scalar or vector valued.

Although the approximation of the gradient seems to be quite rough, experiments show that it leads to promising results. Moreover, it can hardly be surpassed in speed by other methods demanding additional computations. More sophisticated approximation schemes will be implemented and tested in the future, but we expect the gain in accuracy hardly to be worth the loss in efficiency.

Curvature estimation Triangulated surfaces are non-regular surfaces. Therefore, their curvature is not defined everywhere in the sense of differential geometry. To avoid this problem we use either simple heuristics or the rather expensive approximation of the triangulated surface by a smooth surface whose curvature can be calculated.

In the 50's and 60's, the Russian mathematician Aleksandrov [3] developed a theory of non-regular surfaces, in particular, of polyhedral surfaces. The polyhedral metrics they employed to define analogues of Gaussian and mean curvatures were used by Alboul and van Damme [2] for the reconstruction of surfaces from scattered data, recently. See their report [1] for a compact survey on the basic

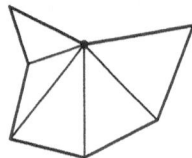 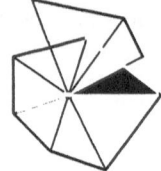 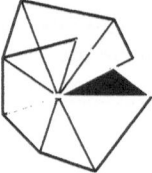

Figure2. Star of a boundary vertex (left) and a non-manifold vertex(middle and right)

ideas of Aleksandrov, and the monograph [4] for a detailed explanation. It turns out that the Gaussian curvature of a vertex v in a non-regular surface with polyhedral metric can be calculated very easily. Let β_i be the inner angle of the triangle (v_i, v, v_{i-1}) which is next to v (see Fig. 1). Then the total angle $\Theta(v)$ in v is defined as

$$\Theta(v) = \sum_{i=1}^{n} \beta_i .$$

If the point x is not a vertex of the triangulated surface, we set $\Theta(x) = 2\pi$. Then, for any point x on the surface the curvature is defined as

$$\text{curv}(x) = 2\pi - \Theta(x) . \tag{3}$$

This expression is also known as the angle deficit. Note that for vertices only we can have a non-zero curvature. In boundary vertices, the curvature is defined as

$$\text{curv}(x) = \pi - \Theta(x) . \tag{4}$$

Besides being mathematically sound, this geometric criterion can be calculated efficiently.

2.3 Topology preserving tests

All vertices are treated equally, assigning them a weight according to (1). But, when a vertex (an edge) is extracted from the priority queue, we perform some tests to preserve feature edges and the topology of the surface. Essentially, it suffices to look for two basic properties of the vertex to be removed, or the vertices adjacent to the edge to be removed, respectively:

- Does the vertex lie on a boundary or a feature edge?
- Is the manifold property satisfied?

Vertices lying on the boundary or on a feature are removed only in the case that the boundary (the feature edge) is continuing in the same direction. This is in some sense equivalent to Schroeders [17] distance to edge criterion, with the difference that for us it is merely an additional test, not the only criterion. Note that the boundary includes also the boundary of interior holes of the surface.

Vertices whose star, i. e. the set of all triangles sharing the vertex, does not have a closed boundary polygon, are assumed to be boundary vertices. If the boundary polygon consists of non-connected components or contains multiple edges outgoing from one vertex (see Fig. 2), the vertex is a complex one. In other words, the surface does not satisfy the 2-manifold property in the neighborhood of this vertex, locally. Complex vertices we do not remove, thus preserving splits of the surface.

3 Complexity of the algorithm

For the computation of the weights as well as for the topology preserving tests, we need the star of a vertex, i. e. the set of all triangles sharing that vertex. Therefore, we store the stars of all vertices in a special array, updating it after each iteration step. The computation of this is linear in the number of vertices in the surface. The priority queue is organized in a heap structure, hence its construction takes linear time as well. So we have a complexity of $O(N)$ for the preprocessing step, N being the number of vertices in the original surface.

Let us have a look at a single iteration step. Extracting the first element of a priority queue can be done in constant time. The topology preserving tests consist merely in sorting the edges of the boundary polygon of the star, hence they take $O(r^2)$ operations, r being the number of triangles in the star. The main effort in each iteration step is enclosed in the updating procedure. Here, we have to update the weights of all vertices belonging to the boundary polygon of the star, which leads to changes of their positions in the priority queue. Changing position, removing, or inserting an element in the priority queue takes $O(\log N^*)$ operations, N^* being the current length of the priority queue. Hence, each iteration step has a complexity of $O(r \log N^*)$. Usually, fans are not growing very much, so r is bounded by a constant number and can be neglected.

In the case of edge collapsing, the computation of the new vertex to be inserted takes a constant effort. When Guéziec's volume preservation strategy is applied, the effort grows up to $O(r)$, where the constant in the $O(\cdot)$-expression can be quite large. However, the re-triangulation required by the vertex removal strategy takes the most time of those three approaches. Although B. Chazelle has shown that an arbitrary polygon with r vertices can be triangulated in $O(r)$ operations [6], his optimal algorithm is of rather theoretical importance. There is a randomized triangulation algorithm by R. Seidel [18] requiring $O(r \log^* r)$ operations which is almost linear in practice. However, in our implementation we currently use a simpler triangulation algorithm [5] with an $O(r \log r)$ effort.

Usually, we have reduction rates of a factor 5–12. This means that we have to perform $O(N)$ removal operations. So the overall complexity of the algorithm can be estimated by $O(N \log N)$. However, this is probably slightly overestimated in practice, because the length of the priority queue is decreasing linearly with the number of iterations and can be next to nothing at the end of the iteration. Hence in practice, the algorithm behaves almost linear, as illustrates Fig. 3 below. These CPU timings were measured on an SGI R10000 (195 MHz).

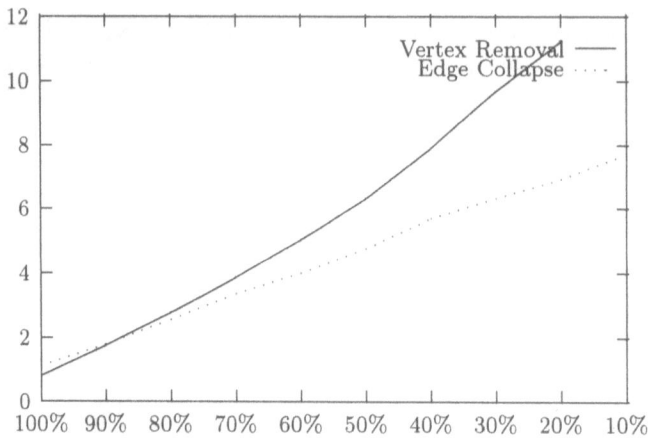

Figure3. CPU timings (in sec) required for the air flow isosurface with 50160 triangles

The example (Fig. 3–5) shows the air flow inside a car cabin. We simplified the isosurface corresponding to a temperature of 18.25°C, where the color represents the pressure inside the car cabin. Since the shape of the surface is not entirely simple, the user-defined parameter α was set to 0.5. The car surface itself was simplified by our geometric surface simplification algorithm. Both methods are integrated into COVISE, the collaborative visualization environment developed by our group (for more detail see [21], [16]).

4 Conclusions

In this paper, we propose a fast algorithm for the data-dependent simplification of arbitrary polygonal surfaces with vertex-attached, possibly multi-dimensional data. First experiments show that the $O(N \log N)$ complexity of the algorithm leads to a almost linear behavior in practice. The edge collapsing approach appears to be still substantially faster than the vertex removal approach due to the comparatively high effort of the re-triangulation procedure. In addition, the triangulation of the simplified surface resulting from the edge collapsing method is often more regular, as can be seen in Fig. 5 and 5. By improving the performance of the re-triangulation these disadvantages will be moderated in future. On the other hand, the vertex removal approach does not cause any interpolation problems for the vertex-attached data, as the edge collapsing does.

Although the proposed gradient criterion is quite rough, it is very efficient and seems to be a sufficiently exact heuristic to mark areas where the data function changes rapidly. The development and comparison of other data-dependent criteria will be subject to further research.

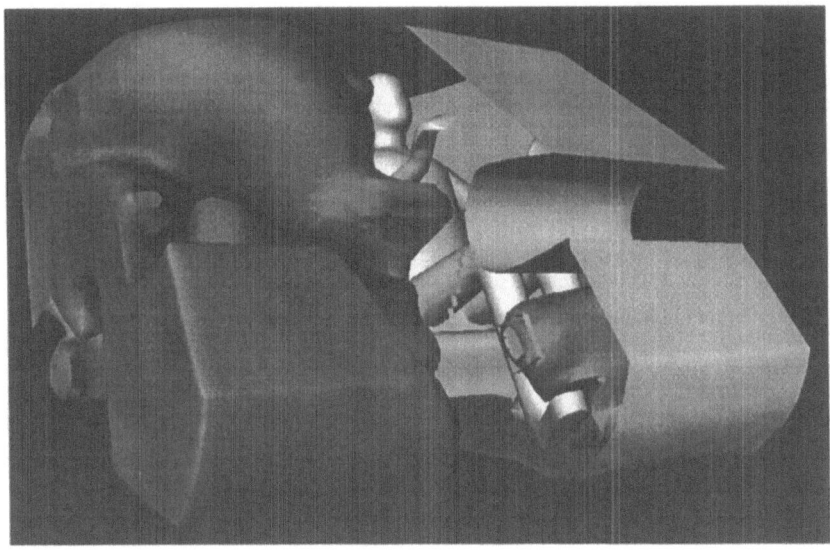

Figure4. The original isosurface (50160 triangles) and car cabin (29104 triangles). The data set is courtesy of Daimler Benz AG.

5 Acknowledgements

During the work on this subject the first author was supported by the European Community in the ESPRIT Project INDEX (EP 22745).

The car cabin air flow data set could be used by courtesy of Daimler Benz, a partner in the INDEX Project.

References

1. L. Alboul and R. van Damme. Polyhedral metrics in surface reconstruction: Tight triangulations. Memorandum No. 1275, Faculty of Applied Mathematics, University of Twente (NL), 1995.
2. L. Alboul and R. van Damme. Polyhedral metrics in surface reconstruction. In *Mathematics on Surfaces VI*, pages 171 – 200. Clarendon Press, 1996.
3. A.D. Aleksandrov. On a class of closed surfaces (in russian). *Mat. Sbornik*, 4:69 – 77, 1938.
4. A.D. Aleksandrov and V.A. Zalgaller. *Intrinsic Geometry of Surfaces*. AMS, 1967.
5. M.de Berg, M.van Krefeld, M. Overmars, and O. Schwarzkopf. *Computational Geometry*. Springer, 1997.
6. B. Chazelle. Triangulating a simple polygon in linear time. *Discrete Comput. Geom.*, 6:485 – 524, 1991.
7. N. Dyn and S. Rippa. Data dependent triangulations for scattered data interpolation and finite element approximation. *Applied Numerical Mathematics*, 12:89 – 105, 1993.

12

8. M. Garland and Heckbert P.S. Surface simplification using quadric error metrics. In *Computer Graphics*, 1997.

9. R. Grosso, Ch. Lürig, and T. Ertl. The multilevel finite element method for adaptive mesh optimization and visualization of volume data. In *IEEE Visualization '97*, pages 387 – 394. IEEE, 1997.

10. A. Gueziec. Surface simplification inside a tolerance volume. Technical Report RC 20440, IBM Research Report, 1997.

11. B. Hamann. Curvature approximation of 3d manifolds in 4d space. *CAGD*, 11:621 – 632, 1994.

12. B. Hamann. A data reduction scheme for triangulated surfaces. *CAGD*, 11:197 – 214, 1994.

13. H. Hoppe, T. DeRose, T. Duchamp, J. McDonald, and W. Stuetzle. Mesh optimization. *Computer Graphics (Proc. SIGGRAPH)*, 27(2):19 – 25, 1993.

14. R. Klein, G. Liebich, and W. Strasser. Mesh reduction with error control. In *IEEE Visualization '96*, pages 311 – 318. IEEE, 1996.

15. J. Popovic and H. Hoppe. Progressive simplicial complexes. *Computer Graphics (Proc. SIGGRAPH)*, 31:217 – 224, 1997.

16. D. Rantzau, U. Lang. A Scalable Virtual Environment for Large Scale Scientific Data Analysis. In: *Proceedings of the Euro VR Mini Conference 97*, Amsterdam, 10.-11. Nov. 1997, Elsevier 1998.

17. W. Schroeder, J. Zarge, and W.E. Lorensen. Decimation of triangle meshes. *Computer Graphics (Proc. SIGGRAPH)*, 26(2):65 – 70, 1992.

18. R. Seidel. A simple and fast incremental randomized algorithm for computing trapezoidal decompositions and for triangulating polygons. *Comput. Geom. Theory Appl.*, 1:51 – 64, 1991.

19. S. Stead. Estimation of gradients from scattered data. *Rocky Mountain J. Math.*, 14(1):265 – 279, 1984.

20. G. Taubin. Estimating the tensor of curvature of a surface from a polyhedral approximation. In *5th Intern. Conf. on Comp. Vision, Los Alamitos*. IEEE Comp. Society Press, 1995.

21. A. Wierse, U. Lang, R. Rühle. Architectures of Distributed Visualization Systems and their Enhancements. *Eurographics Workshop on Visualization in Scientific Computing*, Abingdon, 1993.

22. J. Wilhelms and A. Van Gelder. Multi-dimensional trees for controlled volume rendering and compression. In *1994 Symposium on Volume Visualization*. ACM SIGGRAPH, 1994.

Editor's Note: see Appendix, p. 141 for colored figures of this paper

Data Compression of Multiresolution Surfaces

Reinhard Klein, Stefan Gumhold

WSI/GRIS University of Tübingen, 72076 Tübingen, Germany

Abstract. In this paper we introduce a new compressed representation for multiresolution models (MRM) of triangulated surfaces of 3D-objects. Associated with the representation we present compression and decompression algorithms. Our representation allows us to extract the surface at variable resolution in time linear in the output size. It applies to MRMs generated by different simplification algorithms like local vertex deletion or edge and triangle collapse. The time required to transmit models over communication lines and the space needed to store the MRMs is significantly reduced.

1 Introduction and previous work

Triangle meshes are one of the most popular representations of surfaces for computer graphics applications. On the one hand, rendering of triangles is widely supported by hardware and, therefore, fast. On the other hand, there is an increasing set of data acquisition techniques which generate triangle meshes as output. However, most of these techniques generate much more triangles than necessary to represent the given object with a small approximation error. Isosurface generation can create 1-10 millions of polygons. A digital map of Germany with a resolution of 40 meters in North-to-South and West-to-East direction results in about 500M points. These huge amounts of data lead to problems with data storage and post-processing programs. Animation and real-time rendering of such data sets is almost impossible even on high performance graphics hardware.

Various techniques were published that aimed to reduce surface complexity in order to speed up rendering time [HH92,MSS94,KT96,SZL92,CVM+96,RKH96], [CCMS97,HDD+93,Hop96,RR96,RB93,Tur92,EDD+95]. Aside from the mesh simplification algorithms recent research also focuses on multiresolution representations of triangle meshes. A complex mesh is replaced by several levels of detail (LODs). There are already a number of multiresolution models (MRMs) which make it possible to extract view-dependently simplified meshes at variable resolution [dBD95,dFP95,CPS97], [KLR+95,Pup96,Hop96,KS96].

The models for terrain surfaces and parameterized free-form surfaces in our previous work [KHK96] allow us to refrain from storing the connectivity [1] of the meshes as well as the hierarchy between the different LODS. This leads to a massive data reduction for the MRM. In the other, more general models, the

---

[1] The connectivity of a triangle mesh comprehends all adjacency relationships of the mesh.

connectivity and the hierarchy have to be stored explicitly. Since the number of triangles of the multiresolution model is about three times the number of original triangles, it is often impossible to be keep large models in memory. Data reduction of the connectivity and the hierarchy is indispensable for real-time applications.

A first step into this direction were the progressive meshes proposed by Hoppe [Hop96]. Although with this approach high reduction rates are feasible only restricted selective refinement is possible. His recent paper [Hop97] concentrates on this deficiency but at the expense of storage efficiency. Note that the progressive meshes are based on a special mesh simplification technique, the edge collapse operation. The Multi-Triangulation (MT) introduced by Puppo [Pup96] is more general in the sense that it can be combined with different simplification techniques. In [KK97] we generalized the MT proposed by Puppo for surface meshes embedded in 3D, but for large models the storage needed to store the connectivity and the hierarchy is to high. Therefore, in this paper we propose a new approach, which significantly reduces the storage costs of the model.

2 The simplification algorithm and the multiresolution model

2.1 The simplification algorithm

The simplification algorithm simplifies the input triangulation by successively removing vertices [SZL92]. The minimal one-sided Hausdorff distance between the current triangulation and the simplified one determines which vertex is removed next. A priority queue is used to speed up the computation of the next vertex [KLS96]. All triangles incident to a removed vertex are eliminated from the current triangulation and the resulting hole is retriangulated. From the several possible retriangulations an application dependent optimal one is chosen. A closer look on the different retriangulation strategies reveals that the edge collapse operation can be considered as a special retriangulation technique [KK97]. The simplification algorithm stops if no further vertices can be removed from the simplified triangulation without violating an approximation criterion. The criterion used in our reduction algorithm limits the one-sided Hausdorff distance between the input triangulation and the simplified one.

2.2 The Multi-Triangulation

In the following we briefly describe a data structure for the MT as proposed in [Pup96,KK97]. Besides a list of vertices and a list of triangles the data structure contains a list of *fragments* and a coarse and therefore small starting triangulation T_0, e.g. a tetrahedron. The fragments encode localized refinement steps and consist of two small sets of connected triangles, the coarse *floor* and the finer *ceiling* with the same border polygon, see Figure 1. As the ceiling can replace the floor in order to refine the triangulation, the fragments implicitly define the hierarchy between different levels of detail. In the case of vertex removement, each fragment represents one vertex elimination. The indices of the triangles incident to the removed vertex are listed within the fragment in the list *ceiling* and the indices of the triangles in the retriangulation are stored in the list *floor*. Additionally, the global one-sided Hausdorff distance between the mesh without the removed vertex and the original triangulation is stored in each fragment. Each triangle keeps two pointers to the *upper* and *lower fragment* it belongs to. These interconnect the fragments that overlap and complete the storage of the hierarchy among the fragments.

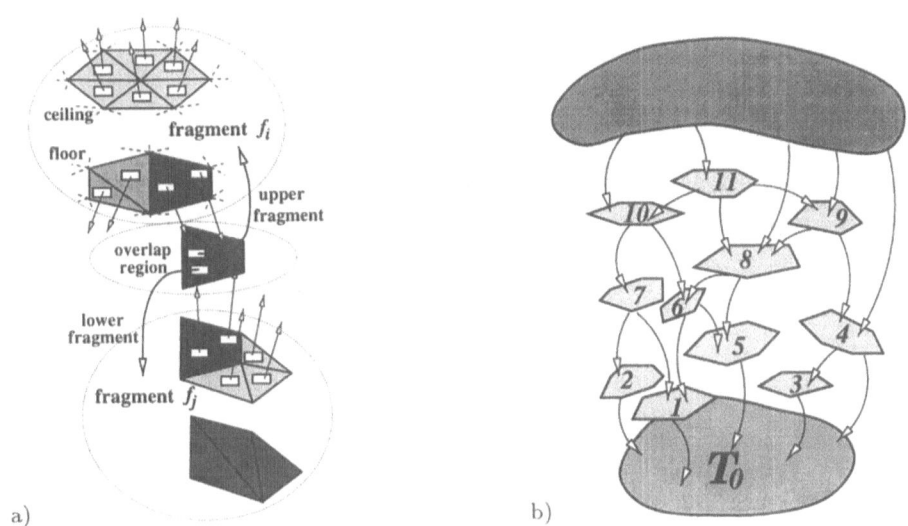

Fig. 1. a) The connectivity of a vertex removement operation is stored in a fragment composed of two lists of triangles – the floor and the ceiling. Each triangle stores its upper and lower fragment, what makes the hierarchy reconstructible. b) The acyclic graph of dependences among the fragments.

3   Using patterns for compression

The main idea of the compression algorithm is based on the following observations:

1. The number of triangles in the fragments is small and bounded by the maximum order of the removed vertices.
2. The fragments with the same number of vertices only differ in the triangulation of the floor.

Therefore, only a few different triangulation types appear in the fragments. For a pentagon all possible triangulations can be obtained from one of its triangulations by rotations. The storage needs for the connectivity of the MT can be significantly reduced if the connectivity information is stored exclusively for the *equivalence classes* [2] of triangulations. To reconstruct a fragment in the current triangulation beside the index of its equivalence class, two indices of adjacent vertices must be known, see figure 2. From these so called *anchor vertices* together with the connectivity of the equivalence class the complete fragment in the current triangulation can be reconstructed, see figure 2. The first anchor vertex determines the position of the fragment in the current triangulation and the second fixes the orientation. In the following the different equivalence classes of triangulations are called *patterns*. In the more general case, where the con-

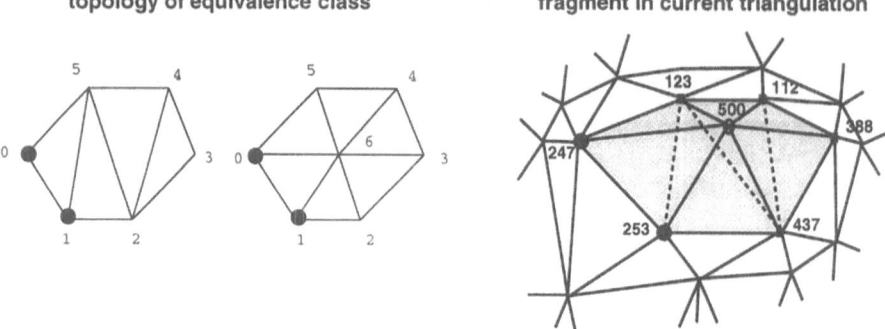

Fig. 2. A fragment is reconstructible from two anchor vertices together with the connectivity of the equivalent class. The anchor vertices 0,1 are equivalent to the vertices 247 and 253 in the current triangulation.

nectivity of the ceiling is not predetermined, the connectivity of the fragment is stored in form of two patterns composing a *rule*.

3.1 Progressive transmission

The model described so far makes progressive transmission possible. For each fragment the index of the corresponding rule, the first anchor vertex and the orientation of the fragment is needed. To store the orientation of the fragment we use the fact that the order in which the vertices are transmitted is fixed. Therefore, among the neighbor vertices of the first anchor vertex exists a vertex with smallest index. Starting from this vertex the neighbors of the first anchor vertex are indexed counterclockwise. If the order of the first anchor vertex is limited to 16, four bits are enough to uniquely define the second anchor vertex and thus the orientation of the fragment. For progressive transmission the storage of each fragment decomposes into one byte for the index of the pattern, four bytes for the index of the first anchor vertex and 4 bits to encode the orientation. This sums up to 44 bits per fragment. In the case of highly reduced models the

---

[2] Each type of triangulation forms an equivalence class.

number of fragments is the same as the number of vertices. This yields nearly the same compression rate as progressive meshes – also for the more general case of vertex removement. The only overhead compared to the progressive meshes is the storage needed for the rules, which is negligible as each floor triangulation can be encoded in one triangle strip and the ceilings are predetermined.

3.2   Compressing the hierarchy

As the hierarchy in this stage is not explicitly stored the model does not allow us to extract a triangulation at variable resolution in time linear in the output size. In the original MT all fragments contain pointers to their triangles and vice versa all triangles contain pointers to their upper and lower fragments to store the hierarchy. Without further compression techniques the amount of storage required is 160 bytes per vertex [3], [Gum98].

Since in our pattern based approach the triangles are implicitly stored in the equivalence classes, the hierarchy is stored using links between the fragments. Suppose for example that the floor of fragment f_i joins some triangles with the ceiling of fragment f_j (compare figure 1). This implies that the ceiling of fragment f_i cannot be inserted before the ceiling of f_j is present in the current triangulation. Therefore, f_i depends upon f_j and we say f_i *directly covers* f_j. In the pattern based approach the hierarchy can be represented by all the direct dependences among the fragments. Therefore, we link for each fragment the directly covered fragments in a closed linked list called *loop* (see figure 3). The crucial point is that the links of each loop are stored within the covered fragments themselves. Thus in each fragment f one downward pointing link and a list of upward pointing links is stored. The downward link begins the loop containing the fragments which are directly covered by f and the list of upward links carry on the loops of the fragments which directly cover f. Each link of a loop consists of a fragment index and a *loop index*. The fragment index specifies the next covered fragment f in the loop and the loop index gives the position of the next link of this loop within the list of upward links of fragment f.

This allows us bi-directional navigation among the directly overlapping fragments. The covering fragments can be found by following all upward links to the end of the corresponding loop. As the number of triangles in a fragment is limited, also the lengths of the loops are limited and therefore the covering fragments can be found in constant time.

To analyze the storage needs of our approach we have to count the number of links in all loops. Each link corresponds to a direct overlap of two fragments. Thus we introduce the *number of overlaps u*. The measurements in section 4 show that the relation between u and the number of fragments f is $u \approx \frac{5}{2} f$ in the case of vertex decimation, which can also be founded with a probabilistic argument [Gum98]. In addition to the u links for each fragment the following quantities have to be stored to give full access to the connectivity and the hierarchy:

1. the indices of two anchor vertices
2. the index of the rule representing the fragment's connectivity

---

[3] We assume that vertex indices and pointers are encoded with 32 bits.

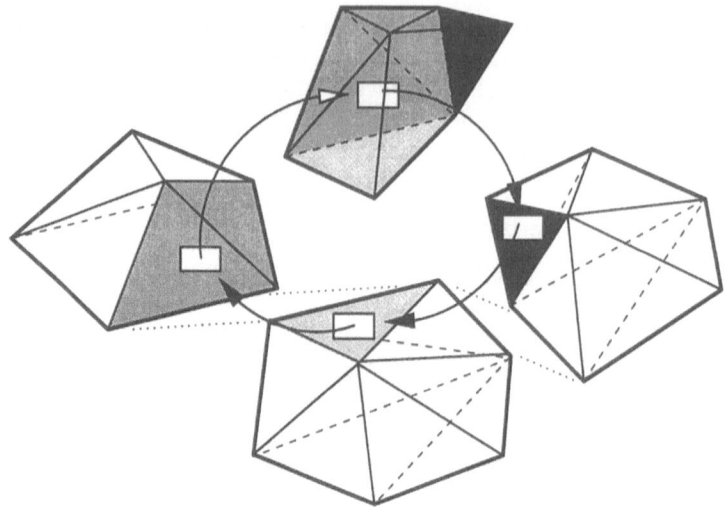

Fig. 3. The floor of the upper fragment covers three other fragments. The four fragments are linked together starting and ending at the covering fragment. Note that the elements of the single linked list are stored in the fragments.

   3. the number of loops the fragment is contained in
   4. the downward link.

As the extracted triangulation is only restricted to the hierarchy, the neighbors of a vertex are not fixed. This is the reason why the second anchor vertex has to be explicitly stored.

In [KK97] it is proposed to limit the order of the removed vertices to accelerate the simplification algorithm. A maximum order of ten limits the maximal number of rules to 7147 [Gum98]. As the number of triangles in the ceiling of the fragments is the same as the order of the removed vertex, also the maximal number of loops a fragment is contained in is limited by ten. Thus the rule index and the number of loops can together be encoded in sixteen bits.

As each loop points back to the beginning, it contains one element more than the number of overlaps it describes. Therefore, the downward links in the fragments must be counted separately. Each link consists of a fragment index and a loop index. If we assume 32 bits per vertex index, also a fragment index can be encoded in 32 bits and the loop index in 4 bits, which sums up to 36 bits per link.

Altogether, the storage needed for the connectivity and the hierarchy in the pattern based multiresolution model sums up to:

$$S_{pbMRM} \approx f \times (64 + 16 + 36)\text{bits} + u \times 36\text{bits}$$
$$\approx f \times \left(116 + \frac{5}{2} \cdot 36\right) \text{bits} = 206f\text{bits}$$

Recalling that the number of fragments is about the number of vertices, 26 bytes per vertex are required for the pattern based multiresolution model. Compared to the MT data structure this corresponds to a reduction factor of six. Again the overhead needed to store the rules is negligible for large models.

3.3 Building a compressed multiresolution model

There are two possibilities to build a compressed multiresolution model. Either the model is extracted from a MT or it is directly build from the simplification algorithm. The first method is slightly simpler as the hierarchy is explicitly stored in the MT. The simplification algorithm successively provides the triangles in the ceiling and the floor of the fragment corresponding to the removed vertex. If for each triangle in the current triangulation temporarily its upper fragment is stored, the covering fragments can be determined from the triangles in the ceiling. The newly produced fragment must be inserted into the loops of the covering fragments.

In both methods the difficult part is to extract the rules from the fragments. Therefore, a comparison algorithm for patterns was developed, which runs in $O(n \times o_{max}^2)$ time, where n is the number of vertices in the pattern and o_{max} the maximal order of its vertices. If the comparison of two patterns is successful, the algorithm produces a map between the pattern vertices. This makes it possible to determine the anchor vertices of fragments whose rule was extracted earlier. As o_{max} is limited to eight [4] and n to ten, the comparison algorithm runs in constant time. Nevertheless to improve performance we defined a Hash-Code on patterns. Major increments are the number of vertices and the vertex orders. To achieve a distinct Hash-Code for patterns and their mirror images the order of the border vertices is exploited. Our Hash-Code is unique for all floor patterns with less than nine vertices [Gum98].

3.4 Incremental selective refinement algorithm

Within a specific application, a criterion must be available to define the required resolution of the multiresolution model. The standard approach is to define a boolean condition c on the triangles of the multiresolution model. If a triangle is invalid, the triangles in the ceiling of its upper fragment are inserted to refine the triangle. In the pattern based approach the triangles are not stored explicitly. Here the condition c is defined on each overlap of two fragments, i.e. on each upward link. This is completely sufficient and if the conditions of the triangles in an overlap can be combined to one condition, the storage space for c is reduced. For example the Hausdorff distance between a triangle and the original mesh can easily be combined by choosing the maximal distance among the triangles in an overlap. Often it is sufficient to define the condition c only on the fragments.

The extracted triangulation additionally must obey to the criterion of "minimal refinement", i.e. it must be the coarsest triangulation that is extractable

---

[4] Only the floor patterns must be compared. These only contain border vertices, the number of which is restricted by the removed vertex.

from the multiresolution model satisfying c. Our incremental algorithm is based on the storage of the current triangulation, which satisfies c and the "minimal refinement" criterion. If the condition c is changed by the application, the current triangulation is adapted to the new condition c in two steps. First the triangulation is coarsened: top fragments [5] containing exclusively valid upward links are removed from the triangulation, until no such fragments are left over. To remove a fragment from the current triangulation, the vertices of the rule are mapped to the vertices of the current triangulation by using the two anchor vertices and a modified comparison algorithm. Then the triangles in the ceiling of the fragment are replaced by the triangles in the floor.

In the second step the current triangulation is refined until all overlaps contained in the current triangulation satisfy the condition c. Refinement is achieved by inserting a fragment: in the current triangulation the floor triangles of a fragment are replaced with its ceiling triangles, introducing the previously removed vertex. Two problems have to be solved:

1. How to find the fragments to the overlaps contained in the current triangulation ?
2. How to ensure that the floor of a to be inserted fragment is part of the current triangulation ?

To solve the first problem, we enumerate the vertices and their corresponding fragments in the order the vertices are removed by the reduction algorithm. This gives us a one to one correspondence between the removed vertices and the fragments. By the way, because of this trick we don't have to store the indices of the removed vertices in the fragments, as they are given by the fragment index. Each triangle in the current triangulation belongs to an overlap that has to be checked for refinement. From the fragments corresponding to the corner vertices of the triangles inserted at last – thus the one with the smallest index – must contain the triangle in its ceiling, because all triangles around an inserted vertex are newly introduced to the current triangulation.

The second problem can be solved with the recursive algorithm *layFoundation*:

Algorithm layFoundation(f)

for all fragments f_i in loop below of f **do**
    if f_i is not inserted **then**
        layFoundation(f_i)
        insertFragment(f_i)
    endif
endfor

Before insertion a fragment must be *founded*, i.e. all triangles in the floor of the fragment are forced into the current triangulation. This is achieved by following

---

[5] A fragment is a top fragment of the current triangulation, if all triangles of its ceiling are contained in the current triangulation. Then the fragment lays on top of the hierarchy restricted to the fragments with ceiling triangles in the current triangulation.

the downward loop to all covered fragments and insert them in the same way. As each fragment is visited exactly once by *layFoundation*, the refinement algorithm is linear in the output size.

4 Measurements

In this section we present measurements on several models of different size. The basic characteristics and the storage needs for the pattern based multiresolution model and the Multi-Triangulation are tabulated in table 1. The implicit surface was produced with a marching cube algorithm. The femur- and the jaw-bones where scanned with a CT-scanner and their triangulations were also obtained by a marching cube algorithm. The face was scanned with a 3D-scanner and triangulated. Satellite elevation data forms the landscape model. All multiresolution models were produced by a simplification algorithm based on vertex decimation.

model	n	f	u	S_{pbMRM}	S_{MT}	$S_{geometry}$	$\frac{S_{pbMRM}}{S_{MT}}$
surface	1340	1303	$2.31f$	36kB	178kB	31kB	4.9
femur	5594	5296	$2.40f$	155kB	795kB	131kB	5.3
face	12530	12446	$2.40f$	346kB	1770kB	294kB	5.2
jaw	12349	11990	$2.39f$	335kB	1759kB	289kB	5.3
landscape	29141	28995	$2.46f$	811kB	4261kB	683kB	5.4

Table 1. The table shows for several models the number of vertices n, the number of fragments f produced by vertex removement, the storage needs for connectivity and hierarchy in the cases of the pattern based model and the Multi-Triangulation, the additional storage needs for coordinates and normals and finally the fraction between the storage needs for connectivity and hierarchy.

Table 1 shows that the storage needs for connectivity and hierarchy in the case of the pattern based multiresolution model is in the order of the storage needs for the geometry, whereas the Multi-Triangulation consumes more than five times more for connectivity and hierarchy. The factor of six between the pattern based approach and the Multi-Triangulation is not quite achieved. The first reason is the storage needs for the rules, which influences especially the smaller models. The second reason is that the models are not completely reduced.

The pattern based approach makes only sense if the time performance is still acceptable for real-time applications. A comparison of the selective refinement algorithms[6] for the pattern based approach and the Multi-Triangulation show that the assymptotic running time is the same. As shown in [Pup96] for vertex decimation with limited fragment size, the running time is linear in the size of the output triangulation. To validate the linear running time for the pattern based approach we measured the time performance of our refinement and

---

[6] The selective refinement algorithm refines the axiom until a demanded resolution criterion is fulfilled.

coarsening algorithm with a view-independent resolution criterion. We started with the axiom, then set the allowed global error to zero and refined to the original triangulation. Subsequently, we set the allowed error to infinity and coarsened the current triangulation down to the axiom. We measured the time needed for several repetitions of this process and computed the average number of vertices which can be removed or inserted per second. It turned out that our implementation of the pattern based approach is only slightly (a factor of 1.5) slower than the corresponding process for our own [KLS96] implementation of the Multi-Triangulation. Table 2 tabulates the measurements for each model.

vert./sec	MHz	cache	second	surface	femur	face	jaw	landsc.
size/KB				36	·155	346	335	811
R4000	100	8 KB	1 MB	14,320	10,787	8,884	8,677	7,890
R4600	100	16 KB	–	15,629	11,509	9,978	9,547	9,613
R8000	75	16 KB	2 MB	23,515	21,900	21,704	20,890	18,222
R10000	175	32 KB	1 MB	60,325	36,689	24,075	23,674	17,481

Table 2. The table shows for each model the number of vertices, i.e. the number of fragments, which can be inserted or removed per second. The measurements were performed on different SGI workstations with different cache sizes.

The measurements were performed on different SGI workstations with different cache and secondary cache sizes. The necessary main memory increases from $67KB$ for the implicit surface to $1.5MB$ for the landscape model. In addition to the model the current triangulation has to be stored during the refinement and coarsening process. In our implementation the representation of the current triangulation in the highest resolution consumes about 1.7 times the storage needs of the connectivity and hierarchy of the pattern based multiresolution model. Together with the geometry data the consumed main memory reaches from $130KB$ to $2.9MB$. The model of the jaw-bone consumes approximately $1MB$.

The R4600 has no secondary cache and the R8000 has enough secondary cache to keep the complete model. On these two work stations the linear time performance is nicely demonstrated. The smaller models are slightly faster because of the primary cache of $16KB$. With a secondary cache size of $1MB$ for the R4000 and the R10000 the analyzed models partially fit and partially do not fit into the secondary cache. Thus the running time is not linear anymore in the size of the model. The larger primary cache of $32KB$ of the R10000 additionally speeds up the algorithms for small models, such that the implicit surface performs 3.5 times faster than the landscape.

In a typical real-time application for selective refinement, there are only a few updates to the current triangulation between two successive frames. Suppose we want 30 frames per second. The refinement and coarsening algorithms can insert or remove about $18,000$ vertices per second on an O2/R10000, thus about 600 vertices per frame. If we further suppose that not more than ten percent of the triangles change between two successive frames and that we may consume about half of the CPU time for the refinement and coarsening algorithm, we can render a simplified scene with about 3,000 triangles in real-time.

5 Conclusion

All simplification algorithms based on vertex removal, edge or triangle collapse are well suited to build up the pattern based multiresolution model proposed in this paper. This multiresolution model supports selective refinement as necessary for view-dependent visualization. The high reduction rates of the storage costs achieved by the use of patterns allow one to process much larger models than with the other multiresolution models described in the literature[7]. We also showed that the time performance of selective refinement is only slightly decreased by the pattern based approach if compared to the Multi-Triangulation.

References

[CCMS97] A. Ciampalini, P. Cignoni, C. Montani, and R. Scopigno. Multiresolution decimation based on global error. *The Visual Computer*, 13(5):228–246, 1997. ISSN 0178-2789.

[CPS97] P. Cignoni, E. Puppo, and R. Scopigno. Representation and visualization of terrain surfaces at variable resolution. *The Visual Computer*, 13(5):199–217, 1997. ISSN 0178-2789.

[CVM+96] Jonathan Cohen, Amitabh Varshney, Dinesh Manocha, Greg Turk, Hans Weber, Pankaj Agarwal, Frederick P. Brooks, Jr., and William Wright. Simplification envelopes. In Holly Rushmeier, editor, *SIGGRAPH 96 Conference Proceedings*, Annual Conference Series, pages 119–128. ACM SIGGRAPH, Addison Wesley, August 1996. held in New Orleans, Louisiana, 04-09 August 1996.

[dBD95] M. de Berg and K. T.G. Dobrindt. On levels of detail in terrains. Technical Report UU-CS-1995-12, Department of Computer Science, Utrecht University, April 1995.

[dFP95] Leila de Floriani and Enrico Puppo. Hierarchical triangulation for multiresolution surface description. *ACM Transactions on Graphics*, 14(4):363–411, October 1995.

[EDD+95] M. Eck, T. DeRose, T. D., H. Hoppe, M. Lounsbery, and W. Stuetzle. Multiresolution analysis of arbitrary meshes. In Robert Cook, editor, *SIGGRAPH 95 Conference Proceedings*, Annual Conference Series, pages 173–182. ACM SIGGRAPH, Addison Wesley, August 1995. held in Los Angeles, California, 06-11 August 1995.

[Gum98] S. Gumhold. Compression of discrete multiresolution models. Technical Report WSI–98–1, Wilhelm-Schickard-Institut für Informatik, University of Tübingen, Germany, January 1998.

[HDD+93] Hugues Hoppe, Tony DeRose, Tom Duchamp, John McDonald, and Werner Stuetzle. Mesh optimization. In James T. Kajiya, editor, *Computer Graphics (SIGGRAPH '93 Proceedings)*, volume 27, pages 19–26, August 1993.

---

[7] The reduction factor compared to Puppo's Multi-Triangulation [Pup96] or Hoppe's Progressive Meshes [Hop97] is about 6.

[HH92]   Charles Hansen and Paul Hinker. Isosurface extraction SIMD architectures. In *Visualization'92*, pages 1–21, oct 1992.

[Hop96]   H. Hoppe. Progressive meshes. In *Computer Graphics Proceedings, Annual Conference Series, 1996 (ACM SIGGRAPH '96 Proceedings)*, pages 99–108, 1996.

[Hop97]   Hugues Hoppe. View-dependent refinement of progressive meshes. In Turner Whitted, editor, *SIGGRAPH 97 Conference Proceedings*, Annual Conference Series, pages 189–198. ACM SIGGRAPH, Addison Wesley, August 1997. ISBN 0-89791-896-7.

[KHK96]   R. Klein, , T. Hüttner, and J. Krämer. Viewing parameter dependent approximation of nurbs-models for fast visualization and animation using a discrete multiresolution representation. In B. Girod, editor, *Herbsttagung '96 3D Bildanalyse und -synthese*, 1996.

[KK97]   R. Klein and J. Krämer. Multiresolution representations for surface meshes. In *Proceedings of the SCCG*, pages 57–66, 1997.

[KLR⁺95] David Koller, Peter Lindstrom, William Ribarsky, Larry F. Hodges, Nick Faust, and Gregory Turner. Virtual gis: A real-time 3d geographic information system. Technical Report 95-14, Graphics, Visualization and Usability Center, Georgia Institute of Technology, USA, 1995.

[KLS96]   R. Klein, G. Liebich, and W. Straßer. Mesh reduction with error control. In R. Yagel, editor, *Visualization 96*. ACM, November 1996.

[KS96]   R. Klein and W. Straßer. Generation of multiresolution models from cad-data for real time rendering. In W. Straßer, R. Klein, and R. Rau, editors, *Theory and Practice of Geometric Modeling)*. Springer-Verlag, 1996.

[KT96]   Alan D. Kalvin and Russel H. Taylor. Superfaces:polygonal mesh simplification with bounded error. *IEEE Computer Graphics and Appl.*, 16(3), May 1996.

[MSS94]   C. Montani, R. Scateni, and R. Scopigno. Discretized marching cubes. In R. D. Bergeron and A. E. Kaufman, editors, *Visualization '94 Proceedings*, pages 281–287. IEEE Computer Society, IEEE Computer Society Press, 1994.

[Pup96]   E. Puppo. Variable reolution of terrain surfaces. In *Proceedings Eight Canadian Conference on Computational Geometry*, August 1996.

[RB93]   J. Rossignac and P. Borrel. Multi-resolution 3d approximation for rendering complex scences. In B. Falcidieno and T. L. Kunii, editors, *Modeling in Computer Graphics: Methods and Applications*, pages 455–465. Springer Verlag, 1993.

[RKH96]   Daniel Cohen-Or Reinhard Klein and Tobias Hüttner. Incremental view-dependent multiresolution triangulation of terrain, 1996. submitted to Pacific Graphics.

[RR96]   R. Ronfard and J. Rossignac. Full-range approximation of triangulated polyhedra. *Computer Graphics Forum*, 15(3):C67–C76, C462, September 1996.

[SZL92]   William J. Schroeder, Jonathan A. Zarge, and William E. Lorensen. Decimation of triangle meshes. In Edwin E. Catmull, editor, *Computer Graphics (SIGGRAPH '92 Proceedings)*, volume 26, pages 65–70, July 1992.

[Tur92]   Greg Turk. Re-tiling polygonal surfaces. In Edwin E. Catmull, editor, *Computer Graphics (SIGGRAPH '92 Proceedings)*, volume 26, pages 55–64, July 1992.

Adaptively Adjusting Marching Cubes Output to Fit A Trilinear Reconstruction Filter

Fabio Allamandri, Paolo Cignoni, Claudio Montani, and Roberto Scopigno

Istituto di Elaborazione dell'Informazione Consiglio Nazionale delle Ricerche,
Via S. Maria, 46 - 56126 Pisa ITALY
Email: {cignoni,montani}@iei.pi.cnr.it , r.scopigno@cnuce.cnr.it

Abstract The paper focuses on the improvement of the quality of iso-surfaces fitted on volume datasets with respect to standard MC solutions. The new solution presented improves the precision in the reconstruction process using an approach based on mesh refinement and driven by the evaluation of the trilinear reconstruction filter. The iso-surface reconstruction process is adaptive, to ensure that the complexity of the fitted mesh will not become excessive. The proposed approach has been tested on many datasets; we discuss the precision of the obtained meshs and report data on fitted meshes complexity and processing times.

1 Introduction

The Marching Cubes (MC) algorithm [11] is nowadays the most diffuse technique for the extraction of iso-surfaces out of volume datasets. The reasons for the MC success include its simple logical structure, implying a nearly straightforward implementation, and its computational efficiency. MC has been incorporated in many commercial and public domain visualization systems. Many papers appeared on enhancements, optimization, extensions and applications of this technique [23, 22, 16, 15, 2, 1, 10]. One of the few limitations of MC is the linearity of the reconstruction kernel used. MC adopts a local approach, i.e. each cell is tested for a possible iso-surface patch independently from the others. Each patch is computed by adopting a table-driven approach, and is defined by the position of vertices located on cell edges. The iso-surface patch returned is therefore a linear approximation (planar faces), whose vertices are located on cell edges (this ensures iso-surface C^0 continuity between cells) and are computed using linear interpolation. When a very high resolution dataset is used, the simplicity of the reconstruction filter is not easily perceptible, unless we perform substantial zooming into the mesh. But if the latter case holds, or if dataset resolution is low, the adoption of a more sophisticated interpolation filter might be required to improve smoothness of the fitted surface.

In this paper we focus on volume data applications based on the visualization of iso-surfaces. We look for methods which produce a "surface-based" output (i.e. ray casting solutions are considered not appropriate), to allow hardware-assisted interactive visualization and data distribution/rendering in web environments.

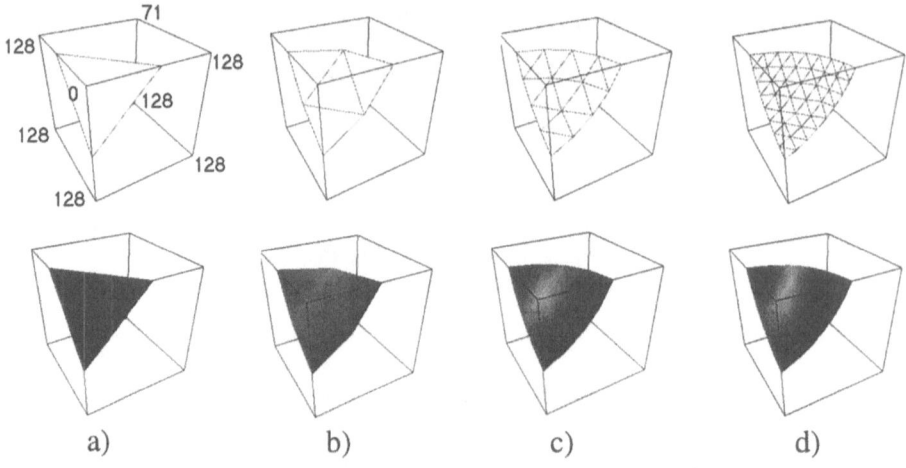

Figure1. Enhancing reconstruction precision: different patches are extracted from the same cell, using increased precision and deeper refinement.

Therefore, we present a solution which improves the precision in the reconstruction process, with respect to standard MC solution, using mesh refinement and the trilinear reconstruction filter (see Figure 1). The iso-surface reconstruction process is adaptive, to ensure that the complexity of the fitted mesh will not become excessive (thus reducing or preventing interactive visualization).

2   Previous work

The excessive simplicity of the reconstruction filter used by MC has been pointed out firstly by Fruehauf [5]. He compared images rendered using the MC output meshes to adopting a ray casting approach (which generally uses a tri-linear reconstruction filter) and showed how much they differ. An advantage of ray casting is to allow the adoption of whichever reconstruction filter; many different interpolation filters have been proposed [12–14] to evaluate/interpolate more precisely both field values and gradients. Unfortunately, ray casting produces images of the isosurface we are interested in (a view-dependent process), rather than extracting explicitly the iso-surface. For many applications, producing an image is not enough. The explicit reconstruction of surface geometries may be needed, for example, in virtual simulation environments. Moreover, a shortcoming of the ray casting approach is the non-interactive rendering time. For these reasons, the precision of the fitted iso-surfaces cannot be improved in many applications by simply adopting a ray casting solution together with a more sophisticated reconstruction filter.

The technique proposed in this paper adopts a regular mesh refinement approach. The idea of improving the quality of a mesh by applying [recursively] a sequence of local refinements is not new, and it has been proposed: to construct adaptive piecewise linear representations of implicit surfaces [6, 21]; to reconstruct adaptively the surface of three-dimensional objects from multiple range images [17]; to extract a surface out of sampled scalar/vectorial 3D datasets starting from an initial surface seed and then applying an iterative surface inflation process [20]; and to refine a surface under a strict surface curvature approximation constraint [9].

The extraction of smooth iso-surfaces has also been recently performed using triangular rational quadratic Bézier patches [7].

3   MC with a trilinear reconstruction filter

The goal is to support a non-linear reconstruction filter in a surface fitting context. The proposed solution has been designed as an extension to the classical MC approach and follows the following list of requirements:

- surface fitting is performed with a *local* approach;
- given a generic reconstruction filter, the simplicial surface mesh produced must approximate the ideal iso-surface defined by the given reconstruction kernel at a user-selected approximation;
- C^0 continuity has to be ensured.

The idea is therefore to enhance the MC algorithm by giving the possibility to refine adaptively each surface patch until the requested precision is fulfilled (Figure 1). The overall pipeline is as follows (with V the volume dataset, q the iso-surface threshold, K the reconstruction filter, ε the given approximation precision, and $maxRec$ the maximum level of recursion which may be produced).

PreciseMC$(V, q, K, \varepsilon, maxRec)$:
    FOR EACH cell $c_{i,j,k} \in V$ DO
        fit an iso-surface patch S on $c_{i,j,k}$ (using standard MC);
        FOR EACH face $f \in S$ DO
            TryToRefine$(f, V, q, K, \varepsilon, maxRec, 1)$;

TryToRefine $(f, V, q, K, \varepsilon, maxRec, lev)$:
    FOR EACH sampling_point p_i on f DO
        evaluate the approximation ε_i of f in p_i with respect to filter K;
        IF $\varepsilon_i > \varepsilon$ THEN $Split\_points := Split\_points + p_i$;
    IF $Split\_points = \{\}$
        THEN output(f)
        ELSE refine f in $\{f_j\}$ (using $Split\_points$);
            FOR EACH f_j DO
                IF $lev \leq maxRec$
                    THEN TryToRefine$(f_j, V, q, K, \varepsilon, maxRec, lev + 1)$
                    ELSE output(f_j);

Mesh refinement is therefore adaptive, because we subdivide only those faces which do not approximate sufficiently the ideal iso-surface. The recursive refinement is halted either if a simplicial approximation which satisfies the given

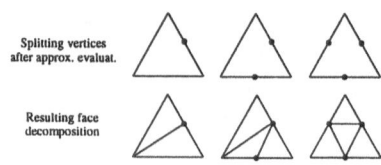

Figure2. Face refinement based on edge midpoints evaluation (rule A).

precision is found, or a maximum recursive level is reached. The user may therefore drive the fitting process by selecting two parameters: the approximation ε and the maximum number of refinement steps $maxRec$.

3.1 Evaluation of approximation and refinement rules

The precision of each face can be evaluated at least in two different manners. A first possibility is to measure a *field-based difference* (i.e. given a simplicial S mesh which approximates the reconstruction filter K, compute the difference between the given threshold value q and the value of the field in the points of S). A second approach is to measure a *geometric difference* between the current iso-surface S and the ideal iso-surface S_K (e.g. evaluate the Hausdorff distance between S and S_K). Both these evaluations may be performed in a precise or in an approximate manner.

We evaluate an approximate *geometric difference* by computing the distance between each face of S and the ideal iso-surface on a discrete number of sampling points. There are many different criteria to select the set of sampling points. A possible choice may be to select the midpoints of the edges (quaternary subdivision). For each of these points p_i, we evaluate the distance between p_i and a corresponding point p'_i on the ideal iso-surface S_K. If this distance is greater than the selected error threshold ε, we classify point p_i as a *splitting point*. Then we refine the current face by inserting the splitting points (the new local triangulation is simply determined by an ad hoc table, see Figure 2). We adopt therefore an heuristic refinement approach, to allow refinement of only a subset of edges. In this case, four different configurations are possible (the three ones represented in Figure 2 plus the one with no splitting vertices). Let us call this refinement criterion Rule A.

Other rules are also possible. One variation of Rule A is to evaluate four splitting points, adding the baricenter to the three edge midpoints (see Figure 3). Let us call it Rule B. Rule B has a disadvantage: because we evaluate all the four split candidates at once, we may decide to split on the central point also when its insertion does not really improve mesh approximation. For example, look at the case when the distance between the central point p_c and the ideal iso-surface S_K is greater than ε, and thus, according to Rule B, we use p_c to split face f. But, if we consider the refinement obtained by using only the other three split points, then in many cases the actual difference between the two refined meshes could

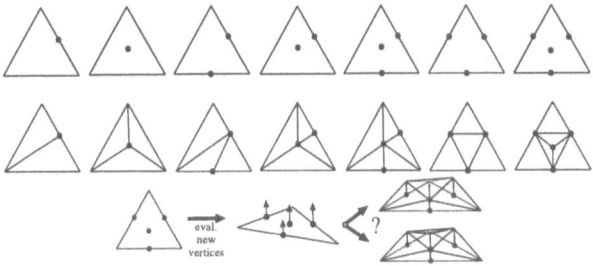

Figure3. Face refinement based on edge midpoint and center point evaluation (rule B); but center point splitting is not always necessary and increases substantially the resulting mesh size.

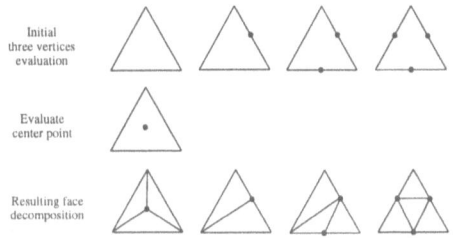

Figure4. Two phases evaluation, under Rule A1.

be much smaller than ε (see Figure 3). In that case we create three new faces that are not really needed to obtain the required approximation. To prevent an excessive increase in the number of faces due to the above reason, we introduce two alternative criteria based on four sampling points. The first one, called Rule A1, extends criterion A by sampling the central point in a second step, only when none of the three edge midpoints is classified as a splitting one (see Figure 4). The second one, Rule A2, always evaluates a fourth sampling point in a second phase. In this case, the initial location of this candidate point is not on the plane of the face to be split, but it depends directly on the current splitting points locations (after relocation on the ideal surface S_K, see Figure 5).

The three different criteria result into different meshes; see the evaluation of the results reported in Section 5.

3.2 Splitting point displacement

In the previous discussion we have not specified how do we find the point p' on the ideal iso-surface S_K which corresponds to the potential splitting point p we are evaluating. A solution is to start a sampling process on the line which originates from the current point p and is parallel to the field gradient in p itself. As far as we sample points on this line, we compute the field value and the current

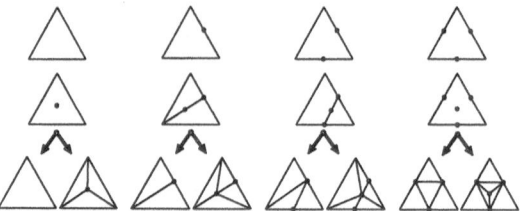

Figure5. Two phases evaluation, under Rule A2.

gradient using the reconstruction filter K. Sampling terminates as soon as we reach the searched value q (i.e. point p'), and we compute the Euclidean distance between p and p'.

Another manner to find point p' can be to analytically compute the nearest intersection between the gradient half-line and the local section of ideal isosurface S_K. This is surely possible in the case of a tri-linear reconstruction filter. But we preferred to adopt the previous solution, based on ray sampling, to be more general: given a reconstruction filter K, we only need to know how to compute K in a generic point p.

The robustness of geometrical computations is obviously a fundamental issue. All of the splitting points are shared between pairs of incident faces. To prevent the occurrence of different values in the replicated evaluation of a candidate splitting point (and potential topological inconsistencies), we must avoid redundant evaluations. All of the evaluated splitting point coordinates (plus accessory information, such as the local geometrical approximation and the gradients computed on such points) are therefore stored in a *hash table*, to prevent redundant evaluations.

4 Management of topologic anomalies

The proposed approach produces an adjustable-precision approximation of the ideal iso-surface by simply refining the standard MC linear patch. This implies that, given a cell and its configuration, we need the initial mesh patch to be topologically correct, otherwise the refinement process can produce erroneous results. Potentially ambiguous configurations of the MC look up table have been identified [4]. These are the configurations where pairs of vertices on a face which are connected by the diagonals have the same classification (both ON or OFF). In general, this problem can be managed with two different approaches. We can adopt a modified MC lookup table [15], which avoids cracks but does not ensure that the surface produced is topologically correct. Or we may directly extract topologically correct geometries, at the expenses of some overhead [18, 16]. The solution proposed by Natarajan [16] chooses for each cell the correct configuration by evaluating the value q' that correspond to a *saddle point* of the local interpolation function.

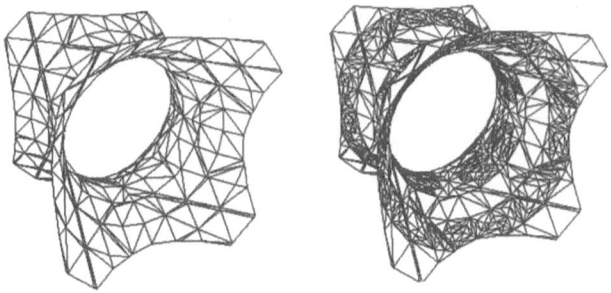

Figure6. Two meshes extracted from the "tube" dataset, using standard MC on the left and the enhanced precision method on the right (max recursion level = 3).

To ensure topological correctness of the initial patches we used the Natarajan's solution , and the overhead introduced resulted small if compared to the processing time spent for the evaluation of the splitting point approximation. But this solution has been defined for the tri-linear reconstruction filter, and how to extend it to the case of a generic filter is not straightforward.

5   Experimental results

The current prototypal implementation of the approach presented, PreciseMC, has been coded in C++. Its GUI has been designed to be used in a distributed client/server environment, using the Java language.
In the current implementation we only provide a trilinear reconstruction filter.
All the times reported have been obtained on a PentiumPro 200Mhz (64 MB RAM). The adoption of an hash table to store splitting points and vertices data (coordinates, approximation error, interpolated normals) has proved highly effective both to improve robustness and to reduce computation times by avoiding redundant computations (in average, times are halved).

PreciseMC has been evaluated on many datasets. We report here results on some iso-surfaces fitted on an 11x14x14 section of the SOD dataset[1], and on two synthetic datasets: "tube" (resolution 9x9x9)[2], and "F" (resolution 30x30x4). Two isosurfaces fitted on the "tube" dataset are shown in Figure 6 (MC on the left, PreciseMC on the right). Figure 7 (see Appendix) shows two meshes extracted from the SOD dataset with threshold 50. The one on the left was fitted using standard MC, and is composed of 654 faces. The mesh on the right was fitted with PreciseMC, using at most five levels of recursion. It is composed of 14,244 faces. Note the difference in the section where the iso-surface bifurcates: the mesh fitted with PreciseMC is much more smoother and thinner.

---

[1] SOD is a regular rectilinear dataset (electron density map of an enzyme), produced by D. McRee, Scripps Clinic, La Jolla (CA).

[2] A sample dataset defined and used in [7].

Table1. Time and complexity of the iso-surfaces fitted on the the "tube" and "F" datasets using the three different splitting points evaluation rules (with 10 the max. number of recursive subdivisions).

"Tube" Dataset (9x9x9, threshold=130.5, MC times = 0.01 sec.)												
precision	Rule A				Rule A1				Rule A2			
(ε)	#faces	maxL	meanL	time	#faces	maxL	meanL	time	#faces	maxL .	meanL	time
1/1000	732	1	0.322	0.050	732	1	0.322	0.070	784	1	0.322	0.081
1/2000	1,256	2	0.782	0.090	1,256	2	0.782	0.130	1,486	2	0.782	0.171
1/4000	1,932	3	1.056	0.160	1,996	5	1.185	0.220	2,732	5	1.084	0.341
1/5000	2,308	4	1.258	0.190	2,420	7	1.484	0.270	3,380	7	1.250	0.420
1/10000	4,412	9	1.968	0.410	4,460	10	2.064	0.551	7,204	9	1.669	0.992
"F" Dataset (30x30x4, threshold=73.5 MC times = 0.01 sec.)												
precision	Rule A				Rule A1				Rule A2			
(ε)	#faces	maxL	meanL	time	#faces	maxL	meanL	time	#faces	maxL	meanL	time
1/100	2,708	7	0.801	0.210	2,726	7	0.805	0.300	3,914	4	0.784	0.471
1/200	4,556	9	1.188	0.411	4,638	10	1.201	0.570	7,726	10	1.205	1.041
1/400	8,350	9	1.558	0.881	8,814	9	1.579	1.142	16,742	10	1.548	2.864
1/800	14,238	10	2.035	1.803	14,450	10	2.040	2.273	32,350	10	2.016	6.830
1/1000	18,058	10	2.186	2.573	18,350	10	2.194	3.095	43,488	10	2.155	10.265

Table2. Time and complexity of the iso-surfaces fitted on the "SOD" dataset with different settings for the approximation precision and the maximum number of recursive decompositions.

SOD dataset (11x14x14, threshold=50.5, MC times = 0.01 sec.)														
	precision = 1/100				precision = 1/500					precision = 1/1000				
maxL	1	2	3	4	1	2	3	4	5	1	2	3	4	5
meanL	0.492	0.674	0.748	0.772	0.843	1.516	1.948	2.138	2.212	0.902	1.689	2.292	2.552	2.843
no. faces	1,141	1,449	1,581	1,603	1,856	3,953	5,925	6,855	7,125	2,074	5,364	9,800	13,080	14,244
time	0.050	0.090	0.110	0.121	0.100	0.240	0.451	0.631	0.701	0.130	0.381	0.841	1.372	1.703

To evaluate the different results produced with the three refinement rules introduced in Section 3.1, we show in Table 1 the size (number of faces, $\#faces$) of the meshes produced by PreciseMC, under the same approximation precision ε. The table reports also the maximum ($maxL$) and mean ($meanL$) recursive depth required to reach the user-selected approximation ε. Precision is given using cell edge size units (e.g. $\varepsilon=1/100$ means precision not less than $1/100$ of the cell edge). From the analysis of these results, we can say that Rule A1 has to be preferred, because it is fast, it is more precise than Rule A and it does not increase too much the size of the mesh. In particular, Rule A2 shows an excessive increase in the size of the mesh produced (nearly the double than the meshes produced with Rules A and A1).

In Table 2 the running times for the SOD dataset with different settings for the approximation precision and the maximum number of recursive decomposition are given.

We measured the actual difference between meshes extracted with MC and PreciseMC using the Metro tool [3]. Metro numerically compares two triangle meshes S_1 and S_2. It performs a surface sampling process on the first mesh, and for each elementary surface parcel it computes a point–to–surface distance with the other mesh. At the end of the sampling process, Metro switches the meshes and execute sampling again, to get a symmetric evaluation of the error.

Metro returns both *numerical* and *visual* evaluations of surface meshes "likeness". We have compared two meshes extracted from the SOD dataset (using the same threshold of Figure 7); the MC mesh is composed of 654 faces, and the PreciseMC one of 14,244. The Metro test gave a maximal distance between the two meshes of 0.39 units (i.e. cell edge length), and a mean distance of 0.12 units. A snapshot of the Metro output is shown in Figure 8; it is zoomed to view in detail the mesh section which describes the thin bifurcation.

6 Conclusions

A new iso-surface fitting solution has been presented, PreciseMC. Given a trilinear reconstruction filter, it improves the precision of the reconstruction process, with respect to standard MC solutions, using an approach based on mesh refinement. The iso-surface reconstruction process is adaptive, to ensure that the complexity of the fitted mesh will not become excessive. Three different refinement rules have been evaluated. Surprisingly, Rule A1 gave the best results; it required low processing times and a reduced increase in the size of the extracted meshes. From a qualitative point of view, the results obtained with PreciseMC are much smoother, more regular and, in some cases, also thinner than those produced with standard MC. PreciseMC shows therefore great potential in medical applications, where it may be selectively adopted to improve the quality of those surfaces which correspond to very thin specimens, such as blood vessels or other internal small cavities. This may improve either the measures taken on the extracted mesh (e.g. to evaluate the occurrence of stenosis or aneurysms in the vessel [19]) or the quality of virtual navigation [8]. Further research is needed to try to extend this approach to other reconstruction filters.

References

1. C.L. Bajaj, V. Pascucci, and D.R. Schikore. Fast isocontouring for improved interactivity. In *1996 IEEE Volume Visualization Symp.*, pages 39–46.
2. P. Cignoni, P. Marino, C. Montani, E. Puppo, and R. Scopigno. Speeding up isosurface extraction using interval trees. *IEEE Trans. on Visualization and Comp. Graph.*, 3(2):158–170, 1997.
3. P. Cignoni, C. Rocchini, and R. Scopigno. Metro: measuring error on simplified surfaces. Technical Report B4-01-01-96, I.E.I. – C.N.R., Pisa, Italy, January 1996.
4. M. J. Dürst. Letters: Additional reference to "Marching Cubes. *ACM Computer Graphics*, 22(4):72–73, 1988.
5. T. Fruhauf. Raycasting opaque isosurfaces in nonregularly gridded CFD data. In P.Zanarini R. Scateni, J.J. van Wijk, editor, *Visualization in Scientific Computing*, pages 45–57. Springer, Wien, 1995.
6. Mark Hall and Joe Warren. Adaptive polygonalization of implicitly defined surfaces. *IEEE CG&A*, 10(11):33–42, 1990.
7. B. Hamann, I.J. Trotts, and G.E. Farin. On approximating contours of the piecewise trilinear interpolant using triangular rational-quadratic Bézier patches. *IEEE ToVCG*, 3(3):215–227, 1997.

8. L. Hong, S. Muraki, A. Kaufman, D. Bartz, and T. He. Virtual voyage: Interactive navigation in the human colon. In *SIGGRAPH 97 Conference Proceedings*, pages 27–34.

9. L. Kobbelt. Discrete fairing. In *Proceedings of the Seventh IMA Conference on the Mathematics of Surfaces*, pages 101–131, 1997.

10. Y. Livnat, H.V. Shen, and C.R. Johnson. A near optimal isosurface extraction algorithm for structured and unstructured grids. *IEEE Trans. on Vis. and Comp. Graph.*, 2(1):73–84, 1996.

11. W.E. Lorensen and H.E. Cline. Marching cubes: A high resolution 3D surface construction algorithm. In *Computer Graphics (SIGGRAPH '87 Proceedings)*, 21(4):163–170, 1987.

12. S. Marschner and R. Lobb. An evaluation of reconstruction filters for volume rendering. In *IEEE Visualization '94*, pages 100–107, 1994.

13. T. Möller, R. Machiraju, K. Mueller, and R. Yagel. Classification and local error estimation of interpolation and derivative filters for volume rendering. In *Proceedings 1996 Symp. on Volume Visualization (Oct. 28-29)*, pages 71–78, 1996.

14. T. Möller, R. Machiraju, K. Mueller, and R. Yagel. A comparison of normal estimation schemes. In *IEEE Visualization '97*, 1997.

15. C. Montani, R. Scateni, and R. Scopigno. A modified look-up table for implicit disambiguation of Marching Cubes. *The Visual Computer*, 10(6):353–355, 1994.

16. B. K. Natarajan. On generating topologically consistent isosurfaces from uniform samples. *Visual Computer*, 11(1):52–62, 1994.

17. Peter J. Neugebauer and Konrad Klein. Adaptive triangulation of objects reconstructed from multiple range images. In *IEEE Visualization '97 - Late-Breaking Hot Topics Session*, October 1997.

18. G.M. Nielson and B. Hamann. The asymptotic decider: removing the ambiguity in marching cubes. In *Visualization '91*, pages 83–91, 1991.

19. A. Puig, D. Tost, and I. Navazo. Interactive cerebral blood vessels exploration system. In *IEEE Visualization '97*, October 1997.

20. L. A. Sadarjoen and F.H. Post. Deformable surface techniques for field visualization. *Computer Graphics Forum*, 16(3):109–116, 1997.

21. L. Velho. Simple and efficient polygonalization of implicit surfaces. *Journal of Graphics Tools*, 1(2):5–24, 1996.

22. J. Wilhelms and A. van Gelder. Topological considerations in isosurface generation. *ACM Computer Graphics*, 24(5):79–86, Nov 1990.

23. Jane Wilhelms and Allen van Gelder. Octrees for faster isosurface generation. *ACM ToG*, 11(3):201–227, July 1992.

Editor's Note: see Appendix, p. 142 for colored figures of this paper

Fast Generation of Multiresolution Surfaces from Contours

Andreas Schilling and Reinhard Klein

Universität Tübingen, Auf der Morgenstelle 10 / C9, 72076 Tübingen, Germany.
E-mail: {andreas—reinhard}@gris.uni-tuebingen.de
http://www.gris.uni-tuebingen.de

Abstract. Surface reconstruction from contours is an important problem especially in medical applications. Other uses include reconstruction from topographic data, or isosurface generation in general. The drawback of existing reconstruction algorithms from contours is, that they are relatively complicated and often have numerical problems. Furthermore, algorithms to generate multiresolution surface models do not exploit the special situation having contours.
In this paper we describe a new robust and fast reconstruction algorithm from contours that delivers a multiresolution surface with controlled distance from the original contours. Supporting selective refinement in areas of interest, this multiresolution model can be handled interactively without giving up accuracy.

1 Introduction and previous work

In medical applications very often tomography-techniques are used to acquire the data. These techniques deliver voxel data sets consisting of a staple of slices (images) where the distance of the slices in general is much larger than the pixel distance within one slice. Therefore, the data can also be considered as an unisotropically sampled voxel set (undersampled in one direction). The general approaches to reconstruct the surface are outlined in Fig. 1. There are two main approaches: a direct and an indirect approach. In the direct approach isosurfaces are extracted from the preprocessed voxel data set using the well known and fast Marching Cubes (MC) algorithm [18]. The preprocessing should perform a continuous classification of the original voxel data, so that the desired surface can be described as an isosurface of the voxel data set. Unfortunately, because of the lack of appropriate alternatives very often only very simple preprocessing like simple noise filtering is applied. If the data set is sampled sufficiently and if the classification doesn't pose problems this approach delivers good results,. If however the sampling in one direction is not sufficient, which is very common e.g. in medical applications, the linear interpolation in the MC algorithm depends on wrong assumptions and the resulting surfaces contains staircase artifacts revealing the slice structure. Furthermore, the resulting surfaces often consist of millions of triangles due to the regular space partitioning used in the MC algorithm. It was soon recognized that without mesh simplification techniques, models produced by the MC could not be handled [26]. As an alternative to the costly simplification algorithms several attempts on adaptive MC schemes with different but not fully satisfactory results [2, 28, 22] have been undertaken.

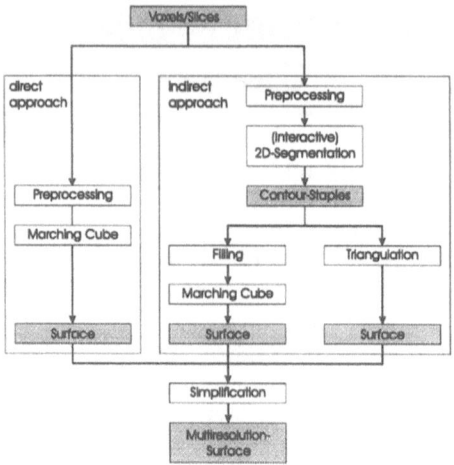

Fig. 1. The main approaches to surface reconstruction from volume data.

In the indirect approach an intermediate step is introduced: contours are extracted from the 2D-slices resulting in a staple of contours. For this purpose 2D-segmentation algorithms are used. In this approach interactive input or manipulation is possible to guide the segmentation. For the reconstruction of the surface from the contour staples again two alternatives are possible, either to voxelize the contours again by filling them and then using a MC algorithm or to construct a triangular facet surface by connecting two neighboring contours. Using the MC algorithm in this setting suffers from the already discussed problems: staircase artifacts (due to the undersampling and 1-bit quantization especially bothersome) and very large surface models. A large number of publications treat the problem of connecting contours, see [14, 8, 4, 17, 3, 20, 1, 23] and [19, 21, 27] for overviews. All these algorithms have to decide which vertices in the neighboring contour must be connected with a given vertex to form triangles. Unfortunately, this correspondence problem cannot be answered uniquely and suffers from the same problem of missing information caused by the undersampling of the original data like the MC algorithm.

1.1 The distance field interpolation

One possibility to overcome the problem of missing data between the slices is to interpolate the so called *distance field* between the contours [17, 24, 10, 7]. Using distance field interpolation, a triangulation of the isosurface can be obtained by applying the MC-algorithm to the unmodified distance-field. In such a way the stair-case artifacts normally produced by the MC-algorithm are avoided. In [13] this idea is realized without mentioning distance-field interpolation. Since our approach is also based on this principle we briefly review this method.

Let Ω be the 3D object and

$$\Omega_i = \{(x,y)|(x,y,z_i) \in \Omega\}$$

a finite set of cross sections. Then the distance fields at the levels $z_0, \ldots, z_n$ are defined
by

$$D_i(x,y) = \begin{cases} -dist((x,y)\partial\Omega_i) & \text{if}(x,y) \in \Omega_i \\ dist((x,y)\partial\Omega_i) & \text{otherwise} \end{cases}$$

where $\partial\Omega_i$ denotes the boundary of Ω_i which is described by the the contours and $dist$
denotes the Euclidean distance within the slices. Now an interpolation of the distance
values in z-direction is used to find intermediate contours (where the interpolated distance is zero). Another approach that can be considered as a special case of the distance
field interpolation is the use of the medial axis between contours from neighboring
slices [23]. The medial axis between the contours of slice i and $i - 1$ is identical to the
contour resulting from a distance field interpolation in the plane $z = 0.5 \cdot (z_{i-1} + z_i)$.
In [23] an approximation of the medial axis is used as the basis of an elaborate reconstruction algorithm that allows to use multiple intermediate levels in cases where the
geometry of subsequent contours is too different. Although in the paper nice results of
this algorithm are shown, a fixed size model is produced. The number of triangles in the
model depends on the number of vertices in the original contours and therefore, on the
accuracy used in the approximation of the contours. The size of the resulting surface
models is smaller than the size of a model produced by the MC algorithm if the contours are approximated with the same accuracy, but not sufficiently small for interactive
rendering. The use of simplification techniques is therefore still necessary.

1.2 Simplification algorithms and multiresolution models

A large number of simplification algorithms for triangle meshes have been developed,
but only a part of them can guarantee a certain geometric approximation error between
simplified and original mesh [16, 6]. However, this error must be known to guarantee
a certain quality for the rendered images of the simplified model. Unfortunately, the
simplification algorithms with this feature are very slow (see [5] for a comparison) or
produce over-estimations of the error [25, 9] that make the results useless for multiresolution models aiming for view-dependent refinement.

In the rest of the paper we describe a new robust and fast reconstruction algorithm
combined with a simplification technique, that exploits the special situation we are
faced with when reconstructing from contours. In this way a very efficient technique
to measure the errors during the simplification process can be used, see 7.1. Section 2
gives a brief overview of the algorithm. Sections 3, 4, and 5 contain the reconstruction
part of the algorithm and section 7 the simplification part.

2 Overview of the algorithm

In principle the algorithm consists of two main steps: the reconstruction of the surface
and the simplification of this surface. However both steps are closely related since the
distance field is used for the reconstruction as well as for the simplification step. The
outline of the algorithm is as follows:

I. Reconstruction **1. Extraction of 2D-contours.** We start with a staple of segmented images, where for each pixel it is known if it belongs to the object or not. In the first step the boundary of the object is determined and the boundary pixels are marked and numbered sequentially. Each contour is identified with a unique number. To avoid topological problems, this step is performed in a grid containing not only the midpoints of the pixels but also their corners. Second, during the extraction of the contours adjacent slices are checked for overlapping areas to identify the connectivity of contours in different slices. **2. Simplification of contours.** After this, each contour is simplified up to a certain approximation error (we use half a pixel). In this way to every edge of a simplified contour the maximum geometric approximation error between the edge itself and the corresponding part of the original contour is known.

3. Computation of medial axes including correspondence. If we consider the simplified contour at the lowest resolution guaranteeing a maximum error of half a pixel, the pixels of the original contour can be regarded as one possible representation in image space and the original pixels can be classified as vertices or inner points of edges. Now, the distance field is computed and afterwards the medial axes between contours from different slices are extracted. To each pixel of the medial axes two pointers to the closest pixels on both contours are stored, see Fig. 2.

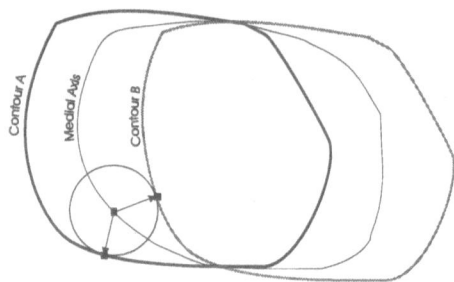

Fig. 2. The arrows from a pixel on the medial axis to the two contours A and B respectively indicate the correspondences to the closest pixels on the contours.

4. Surface triangulation. After the computation of the medial axes and the correspondences above, a triangulation of the resulting surface is computed. The main idea is to trace the medial axes and use the correspondences to pixels on the two neighboring contours to connect the vertices by edges.

II. Simplification **1. Edge collapse.** In the simplification algorithm simple edge collapse operations are performed [12]. The order of the edge collapse operations is determined by a priority queue based on the error that would be introduced if the edge was collapsed. **2. Error measurement.** To measure the error introduced by an edge collapse, the distance field computed above is exploited. The intersection lines of the newly created triangles with the respective slices are rendered (using Bresenham algorithm) into the images of the slices. While rendering the distance values are read from the distance map of the corresponding slice, see Fig. 8.

3 Simplification of contours

To simplify the contours a modified version of the Douglas-Peucker-algorithm is used [11]. This algorithm starts with one arbitrary vertex of the original contour. In each subsequent step a further point (with greatest distance to the current polygon) of the original contour is inserted. To find this point with greatest distance a convex hull technique is used, that reduces the complexity of the algorithm from $O(n^2)$ to $O(n \log n)$.

4 Computation of the medial axes and correspondences

4.1 The distance field

For our purpose the application of a simple distance transform to compute the distance field and the medial axes is not sufficient, since we also want to know for each pixel which is (are) the closest pixels on the border. Therefore, we use a simple filling technique based on the following observations: Starting from a point on a contour there exist only four basic configurations of neighboring contour pixels from which all others can be deduced using symmetries, see Fig. 3. For each configuration the area of pixels that are closer to the pixel than to its two neighbors can easily be determined. Therefore, for each contour pixel only certain image pixels are visited and filled with the distance to the contour pixel.

During the filling a pointer to the border pixel from which the filling started is stored in each image pixel. The filling stops if the border of the image is reached or lower distance values are already present in a pixel, that indicate that further pixels to be filled are closer to another part of the contours of the same slice.

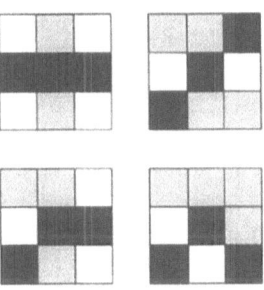

To speed up the filling process and reduce the number of updates of distance values in a pixel we first fill horizontal and vertical lines and only then the more complicated areas. A further speed up is achieved by starting the filling on border pixels that belong to a coarse rectangular grid, thereby considerably reducing the areas to be filled.

Fig. 3. Four basic configuratuions of contour pixels. Black: Contour pixels, gray: pixels to be filled with distance values, white: pixels, that are filled starting from other contour pixels.

4.2 The medial axis

After the computation of the distance field it is easy to extract the medial axes between contours of neighboring slices. First the two distance fields are added. The resulting image contains connected regions i of positive (including zero) or negative values, respectively, see Figure 4. The borders between these areas constitute the medial axes. These borders are extracted with an algorithm which scans the image for a change of the sign and then immediately traces and marks the pixels of encountered medial axes. For each new medial axis a pointer to one of its pixels is stored.

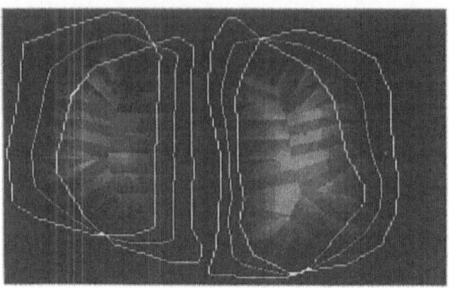

Fig. 4. Adding the two distance fields of consecutive slices delivers the medial axes at pixels where the sign changes. Each slice contains two contours. The gray values in this image encode the distance to the contours as well as the correspondences to the different contour pixels. For a color version of this image see Fig. 10 (see Appendix).

5 Triangulation of the surface

Using the medial axes and the known correspondences to the closest points on the contours the triangulation of the resulting surface is straightforward.

The basic step is to trace the medial axes, which are closed polygons. The tracing starts at an arbitrary pixel of the medial axis. Let M be the medial axis and $M_i, i = 1, \ldots, m$ its pixels. Let P and Q denote the contours corresponding to M and let $P_{i,0}, Q_{k,0}$ be the vertices of the polygon approximating P and Q respectively with a maximum approximation error of $1/2$ pixel. The pixels of the original contours between the vertices are numbered with the second index e.g. the p_i pixels $P_{i,1}, P_{i,2}, \ldots, P_{i,p_i}$ between $P_{i,0}$ and $P_{i+1,0}$, where $P_{i,p_i+1} = P_{i+1,0}$, see Fig. 5.

When the pixels of the medial axis are traced in a sequential way, for each of its pixels the numbers i, k of the two corresponding contour edges $P_{i,0}P_{i+1,0}$ and $Q_{k,0}Q_{k+1,0}$ are recorded. If, while tracing the medial axis the corresponding pixel on one of the contours jumps to the next edge, e.g. from edge i to edge $i + 1$, the vertex $P_{i+1,0}$ is put

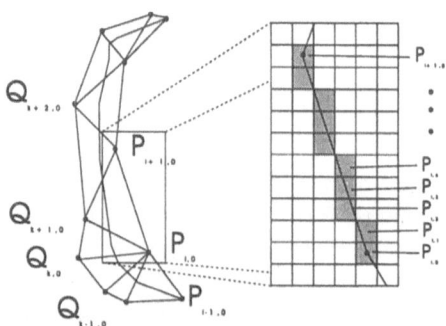

Fig. 5. Triangulation of the contour (see text for details).

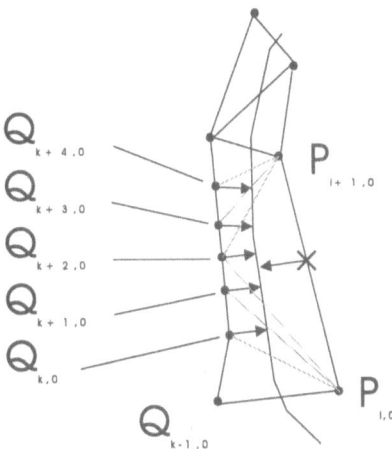

Fig. 6. Vertices $'Q_{k,0}$ through $Q_{k+4,0}$ are connected to $P_{i,0}$ or $P_{i+1,0}$ resp. due to their correspondences on the medial axis.

onto a stack S^1. At the beginning we trace until in the sequence of recorded vertices on contour P is followed by a vertex on contour Q or reverse. For simplicity we assume for the following that we had a change from P to Q. Then the two vertices are connected by an edge and all other vertices are removed from the stack and the tracing proceeds until again a change in the recorded vertices, now from contour Q to P occurs. Again these two vertices are immediately connected and all remaining vertices that are already on the stack are processed in the following way. The stack contains now a sequence of the following form $PQ \ldots QP$, with one or more vertices from contour Q between two vertices of contour P. The polygon defined by these vertices can be triangulated in different ways. Nice triangulations can be achieved exploiting the correspondences in the following way. We connect all vertices on Q that correspond to pixels on the medial axis that are closer to $P_{i,0}$ than to $P_{i+1,0}$ with $P_{i,0}$ and the others with $P_{i+1,0}$, see Fig. 6.

Until now we have assumed that we always found consecutive edges while tracing the contours, but sometimes this may not happen, see Fig. 7. In these cases new vertices are inserted into the contours. Let us assume that the correspondence changes as shown in Fig. 7. Then we insert the vertices $P_{i,\alpha}, P_{m,\beta}, Q_{j\delta}$ into the contours P and Q, respectively and we introduce one of the two pixels on the medial axis M_i or M_{i+1} (we choose M_i). The corresponding stripe is triangulated as shown in Fig. 7 and the vertices $P_{m,\beta}$ and $Q_{j,\delta}$ are put on the stack[2]. Now the algorithm can proceed as described above until all medial axes have been processed.

After processing all medial axes there remain parts of contours that had no correspondence to a medial axis and are therefore not fully integrated in the triangulation.

---

[1] The handling of the special (simpler) case, where P and Q corresponds to the same pixel on the medial axis is handled is not described here, but is straight forward

[2] In some cases better triangulations can achieved with a more elaborated algorithm that inserts two additional vertices on the medial axis into the triangulation.

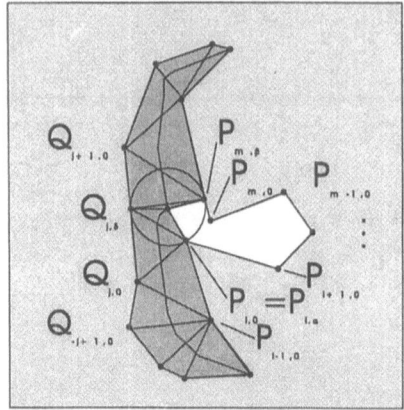 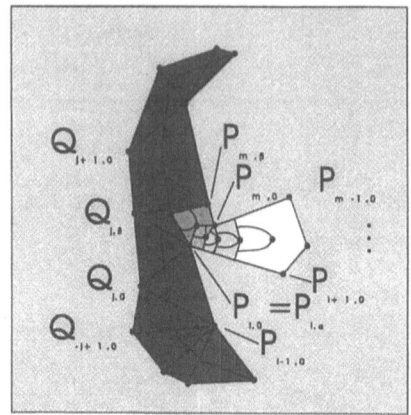

Fig. 7. Left: When edges without correspondence on the medial axis are present, additional vertices are introduced. Then the dark gray area can be triangulated. The white area remains to be handled as shown on the right side: Various shades of gray show areas gained by multiple recursive computations of medial axes between the vertices inserted on the medial axes and the remaining part of the contours. In the Figure six recursion step are performed. The last inserted vertex on the midline lies on a level of only $\frac{1}{32}$ of the distance between the slices above the lower slice. Therefore, the recursion could as well be terminated earlier without giving up much of the quality.

These areas could be triangulated resulting in flat areas parallel to the slices (with the exception of the vertices on the medial axis in between the two slices) [1]. But to improve the results the newly introduced vertices on the medial axes can be considered as new contours (in between the original contours) and the described algorithm (including the calculation of new distance fields in the respective areas) is applied recursively, see Fig. 7. Of course, contours without a neighboring contour in the next slice (end of the object, end of the contour staple) are closed by a plane triangulation in the slice of this contour. During simplification these triangles have to be treated separately in such a way that the distance between the original position and the simplified triangulation can be controlled.

Note, that the results achieved with our simple and fast triangulation algorithm are similar to the ones achieved with the algorithm of Oliva et al [23].

6 Review of multiresolution models

6.1 Generating the multiresolution model

The generation of a MRM of an object generally involves a sequence of local simplification operations like vertex removal, edge collapse, triangle collapse or vertex clustering. The sequence of local simplification operations defines a sequence of coarser and coarser approximations of the original model, the MRM. How this sequence is generated depends on the various simplification algorithms. In general a mesh simplification algorithm starts with the finest triangulation in 3D space approximating the original model. Then it simplifies the starting triangulation by clustering vertices, by collapsing edges or triangles or by removing vertices from the current triangulation and retriangulating the resulting holes. This is done until no further simplification step can be performed. In many algorithms the order in which the simplification steps are performed is determined by a priority queue. A cost function is evaluated for each possible simplification operation and the one with the lowest cost is performed. In general the cost function represents the error (geometric distance) between original and simplified mesh.

6.2 Selective refinement of multiresolution models

If the inverse local simplification operations are known (e.g. vertex split as the inverse of edge collapse operation), we are able to refine a coarse approximation of the model by reversing the whole simplification process. However, if we want to perform only selective refinement we have to find a way to skip parts of the inverse simplification process and thereby change the sequence of refinement operations. Of course this is not arbitrarily possible (e.g. we cannot split a vertex which is not present in the current mesh). The dependencies between the different simplification steps define a hierarchy that can be described by a directed acyclic graph of modification operations or the associated triangles. Therefore, a general selective refinement algorithm starts with a crude approximation of the model and checks for each triangle if refinement is needed. If yes, the algorithm has to take care that all predecessor operations of the needed refinement operation have already been performed. The next section describes the measure that can be used to decide about the need of further refinement of a certain triangle.

7 Simplification

The most expensive part of the simplification algorithm is the evaluation of the cost function. In our case we need a geometric distance between the original contours and the simplified model. According to our experience the quality of the resulting triangulation does mainly depend on the order of the different simplification steps and not on the special topological operations like vertex removal, edge- or triangle collapse [15]. Therefore, we use a simple edge collapse technique, where no new vertices are introduced.

7.1 Measuring the error

To evaluate the cost function, that is to measure the error between original and simplified surface model we use the fact that in each slice the distance field is already calculated and delivers automatically a set of envelopes of the original contour. Based on this observation the measurement of an error that would be introduced if an edge was collapsed can easily be performed in the following way: Let $\Delta = \Delta(p_j, p_k, p_l)$ be a triangle generated if the edge was collapsed spanning the slices $m, m+1, \ldots, n-1, n$. Consider the line segments l_i, $m \leq i \leq n$ defined as the intersection between Δ and the planes $z = z_i$. For each slice we want to find the maximum distance between the

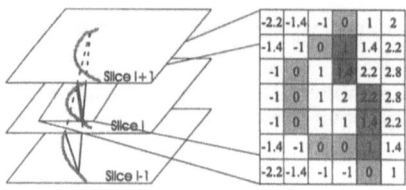

Fig. 8. Measuring the approximation error by tracing Bresenham-lines in the distance fields of the slices.

points of l_i and the contour Ω_i. To find this distance we first read the distance field at the pixels intersected by the triangle Δ. These pixels can easily be determined by 'drawing' l_i using a Bresenham algorithm, see Figure 8. [3]

Note, that like in the simplification envelopes algorithm [6] also in our approach a one-sided Hausdorff- distance between original and simplified triangulation is measured. Therefore, if this distance is smaller than a certain ϵ, we guarantee that for every point p of the simplified triangulation we can find a point q on the original triangulation with $d(p, q) \leq \epsilon$, but the inverse relation does not hold.

7.2 Acceleration of distance computation

For errors larger than 2 pixels in image space the read out of the distance values can be accelerated by skipping $\lfloor \epsilon - d - 1 \rfloor$ pixels, where d is the distance readout at the current position and ϵ is the already reached error between original and simplified model, as from pixel to pixel in image space the distance can grow at most by 1. In this way the drawback of the simplification envelopes algorithm [6] of having a fixed envelop and therefore not being able to build up a reasonable multiresolution model is avoided.

---

[3] To obtain a smaller (but still conservative) estimation for the deviation from the original contour, the maximum of the values read out from the distance field could be multiplied with $\sin(\alpha)$, where α is the angle between the normal of Δ and the z-direction, see Fig. 9. However, in this case it is important to take care of the bounds of the triangles.

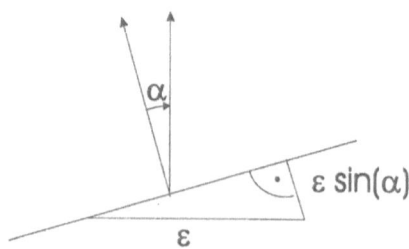

Fig. 9. The error ϵ read out from the distance field is multiplied by $\sin(\alpha)$ to get the real approximation error.

8   Conclusion and further work

The contribution to the problem of reconstruction from contours presented in this paper is a twofold. On one hand the distance field is used for a robust reconstruction algorithm based on the medial axes. In this algorithm the distance field is not only used to calculate the medial axes, but also delivers correspondences used for an excellent triangulation. On the other hand, the second big problem of current reconstruction algorithms, the huge number of resulting triangles, is solved with a new fast simplification algorithm that exploits the (already calculated) distance field to guarantee a certain approximation error between the simplified surface models and the original contours.

This guarantees that in each level of detail, the contours are approximated with a certain approximation error, since the intersections between the slices and the simplified surfaces are within a certain envelop. A problem not yet recognized in the literature is that in general it is not sufficient to guarantee only the distance of the simplified models to certain points or vertices, since it may happen that between two consecutive simplification steps although the distance to the points or vertices remains small or even constant the distance between the two consecutive triangulations can be arbitrarily large. This leads to artifacts in animations where the level of detail is changed. Currently we are working on this problem. Our present research includes also the generalization by using 3D-distance fields to control the approximation error.

9   Acknowledgement

We would like to thank B. Kreher for all the programming efforts which were necessary to gain the described results.

References

1. Gill Barequet and Micha Sharir. Piecewise-linear interpolation between polygonal slices. *Computer Vision and Image Understanding: CVIU*, 63(2):251–272, March 1996.
2. Jules Bloomenthal. Polygonizatino of implicit surfaces. *Computer Aided Geometric Design*, 5(4):341–355, November 1988.
3. Jean-Daniel Boissonnat. Shape reconstruction from planar cross sections. *Computer Vision, Graphics, and Image Processing*, 44(1):1–29, October 1988.
4. H. N. Christiansen and T. W. Sederberg. Conversion of complex contour line definitions into polygonal element mosaics. *Computer Graphics*, 12(3):187–192, August 1978.
5. P. Cignoni, C. Montani, and R. Scopigno. A comparison of mesh simplification algorithms. *Computers & Graphics*, 22, 1998.
6. J. Cohen, A. Varshney, D. Manocha, and G. Turk. Simplification envelopes. *Computer Graphics*, 30(Annual Conference Series):119–128, 1996.
7. Daniel Cohen-Or, David Levin, and Amira Solomovici. Contour blending using warp-guided distance field interpolation. In *IEEE Visualization '96*. IEEE, October 1996. ISBN 0-89791-864-9.
8. H. Fuchs, Z.M. Kedem, and S.P. Uselton. Optimal surface re-construction from planar contours. *Comm. of the ACM*, 20:693–702, 1977.

9. Michael Garland and Paul S. Heckbert. Surface simplification using quadric error metrics. In Turner Whitted, editor, *SIGGRAPH 97 Conference Proceedings*, Annual Conference Series, pages 209–216. ACM SIGGRAPH, Addison Wesley, August 1997. ISBN 0-89791-896-7.

10. Gabor T. Herman, Jingsheng Zheng, and Carolyn A. Bucholtz. Shape-based interpolation. *IEEE Computer Graphics and Applications*, 12(3):69–79, May 1992.

11. John Hershberger and Jack Snoeyink. Speeding up the Douglas-Peucker line-simplification algorithm. In P. Bresnahan et al., editors, *Proc. 5th Intl. Symp. on Spatial Data Handling*, volume 1, pages 134–143, Charleston, SC, August 1992.

12. Hugues Hoppe. Progressive meshes. In Holly Rushmeier, editor, *SIGGRAPH 96 Conference Proceedings*, Annual Conference Series, pages 99–108. ACM SIGGRAPH, Addison Wesley, August 1996. held in New Orleans, Louisiana, 04-09 August 1996.

13. M. W. Jones and Min Chen. A new approach to the construction of surfaces from contour data. *Computer Graphics Forum*, 13(3):C/75–C/84, ???? 1994.

14. E. Keppel. Approximating complex surfaces by triangulation of contour lines. *IBM J. Res. Dev.*, 19:2–11, 1975.

15. R. Klein and J. Krämer. Multiresolution representations for surface meshes. In *Proceedings of the SCCG (Spring Conference on Computer Graphics), Budmerice, Slovakia*, pages 57–66, 1997.

16. Reinhard Klein, Gunther Liebich, and Wolfgang Straßer. Mesh reduction with error control. In *IEEE Visualization '96*. IEEE, October 1996. ISBN 0-89791-864-9.

17. D. Levin. Multidimensional reconstruction by set-valued approximation. *IMA J.Numerical Analysis*, (6):173–184, 1986.

18. W. E. Lorensen and H. E. Cline. Marching cubes: a high resolution 3D surface construction algorithm. In M. C. Stone, editor, *SIGGRAPH '87 Conference Proceedings (Anaheim, CA, July 27–31, 1987)*, pages 163–170. Computer Graphics, Volume 21, Number 4, July 1987.

19. Michael Lounsbery, Charles Loop, Stephen Mann, David Meyers, James Painter, Tony DeRose, and Kenneth Sloan. Testbed for the comparison of parametric surface methods. In L. A. Ferrari and R. J. P. de Figueiredo, editors, *Curves and Surfaces in Computer Vision and Graphics (Proceedings of SPIE)*, volume 1251, pages 94–105, 1990.

20. David Meyers, Shelley Skinner, and Kenneth Sloan. Surfaces from contours. *ACM Transactions on Graphics*, 11(3):228–258, July 1992.

21. Heinrich Müller and A. J. Klingert. Surface interpolation from cross sections. In H. Hagen, H. Mueller, and G. Nielsen, editors, *Focus on Scientific Visualization*, pages 139–190. Springer-Verlag, 1993.

22. Heinrich Müller and Michael Stark. Adaptive generation of surfaces in volume data. *The Visual Computer*, 9:182–199, 1993.

23. J. M. Oliva, M. Perrin, and S. Coquillart. 3D reconstruction of complex polyhedral shapes from contours using a simplified generalized Voronoi diagram. *Computer Graphics Forum*, 15(3):C397–C408, September 1996.

24. Bradley A. Payne and Arthur W. Toga. Distance field manipulation of surface models. *IEEE Computer Graphics and Applications*, 12(1):65–71, January 1992.

25. R. Ronfard and J. Rossignac. Full-range approximation of triangulated polyhedra. *Computer Graphics Forum*, 15(3):C67–C76, C462, September 1996.

26. William J. Schroeder, Jonathan A. Zarge, and William E. Lorensen. Decimation of triangle meshes. *Computer Graphics*, 26(2):65–70, July 1992.

27. L.L. Schumaker. Reconstructing 3d objects from cross-sections. In W. Dahmen, M. Gasca, and C.A. Micchelli, editors, *Computation of Curves and Surfaces*, pages 275–309. Kluwer Academic, Dordrecht/Norwell, MA, 1989.

28. R. Shu, C. Zhou, and M. S. Kankanhalli. Adaptive marching cubes. *The Visual Computer*, 11(4):202–217, 1995. ISSN 0178-2789.

Editor's Note: see Appendix, p. 143 for colored figure of this paper

Part II
Feature Extraction

Experiments on the Accuracy of Feature Extraction

Freek Reinders[1], Hans J.W. Spoelder[2], and Frits H. Post[1]

[1] Dept. of Computer Science, Delft University of Technology
PO Box 356, 2600 AJ Delft, The Netherlands
email: {k.f.j.reinders, f.h.post}@cs.tudelft.nl
[2] Dept. of Physics and Astronomy, Vrije Universiteit
De Boelelaan 1081, 1081 HV Amsterdam, The Netherlands
email: hs@nat.vu.nl

Abstract. Feature extraction is an approach to visualization that extracts important regions or objects of interest algorithmically from large data sets. In our feature extraction process, high-level attributes are calculated for the features, thus resulting in averaged quantitative measures. The usability of these measures depends on their robustness with noise and their dependency on parameters like the density of the grid that is used. In this paper experiments are described to investigate the accuracy and robustness of the feature extraction method. Synthetic data is generated with predefined features, this data is used in the feature extraction procedure, and the obtained attributes of the feature are compared to the input attributes. This has been done for several grid resolutions, for different noise levels, and with different feature extraction parameters. We present the results of the experiments, and also derive a number of guidelines for setting the extraction parameters.

Keywords: feature extraction, attribute calculation, experimental accuracy estimation.

1 Introduction

Feature extraction is a set of techniques in scientific visualization aiming at algorithmic, automated extraction of relevant features from data sets. This leads to a small set of numbers (the attributes) describing the properties of the features. Hence, feature extraction lifts the data to a higher abstraction level, and comes down to a major data reduction. Since an "interesting feature" is different for each application, many application-specific feature extraction techniques exist, examples are critical points extraction [2], vortex extraction [1], and shock wave extraction [3]. A more general approach for extracting features is introduced by Post et al [4], and [8]. It is summarized by the pipeline model in figure 1, and consists of the following stages: selection, clustering, attribute calculation, and iconic mapping.

Selection identifies all grid nodes where the data satisfies a certain selection criterion, *clustering* clusters the selected nodes into regions of interest, *attribute*

50

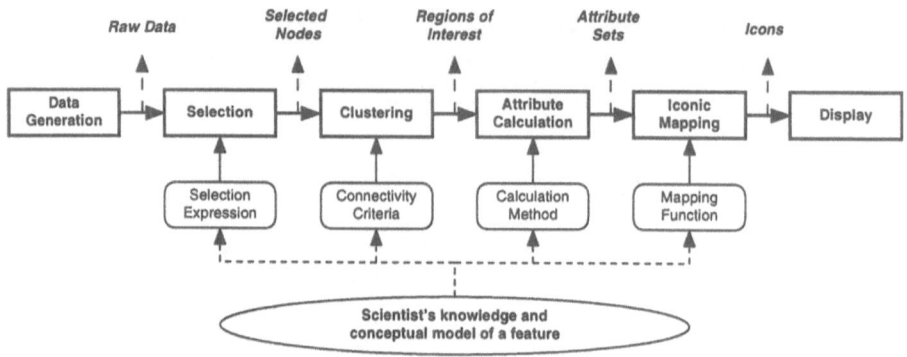

Fig. 1. The feature extraction pipeline.

calculation determines a number of attributes for each feature, and *iconic mapping* maps the calculated attributes to an icon which can be displayed. This process is controlled by the scientist in the sense that his knowledge of the data and his conceptual ideas of an interesting feature are translated into the selection expression, the connectivity criteria, the calculation method and the mapping function.

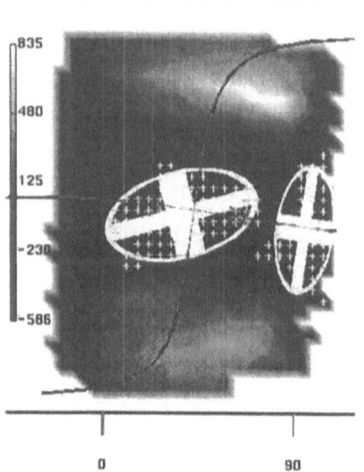

Fig. 2. Iconic presentation of cloud features on Venus.

An example of feature extraction is the detection of cloud formations in the atmosphere of Venus [6]. The clouds are visualized by ellipsoids which give a good indication of position and size (see Fig. 2). Motion of the clouds can be derived by visually matching the ellipsoids in consecutive frames. However, we believe the attributes of the ellipsoids can be used for automatic matching of features. Subsequently, it is important that the attributes are determined accurately, and that the results are robust with respect to noise. The latter will depend on feature extraction parameters like the selection threshold value, the cluster threshold, and the connected component definition.

Therefore, we wish to investigate the accuracy of the attribute calculation, and the influence of noise in combination with differ-
ent extraction parameters on the calculated attributes. This is achieved by a simulation study: synthetic data is generated with synthetic features, i.e. with known attributes, and with noise with a known distribution function. This data is used as input for feature extraction, and the attributes obtained are compared to the initial settings of the attributes. This has been done for different noise levels, with different grid densities, and with different feature extraction param-

eters. In this way, we derived a number of guidelines for working with the feature extraction method in practice. Hence, it is not our intention to extract features below noise level!

The paper is organised as follows: section 2 gives a detailed description of the problem definition, section 3 discusses the generation of the synthetic data, section 4 describes the experiments performed, section 5 presents the results of the experiments, and section 6 draws some conclusions, and finally section 7 suggests work for future research.

2 Problem definition

The experiments focus on two main issues:

1. *Accuracy of the attribute calculation method.* Attribute calculation determines a number of quantitative characteristics of a feature. The attributes may be related to the data in the feature, to the geometry of the feature, or to a combination of both. In order to describe the geometry, we use ellipsoid-fitting because amorphous 3D objects can be approximated by ellipsoids [7]. The resulting attributes are the center position, the lengths and orientations of the ellipsoid axes. These can be estimated using an integration over the selected nodes: the average position of the nodes is the center position of the ellipsoid, and from the variance-covariance matrix of the node positions the axis-lenghts and orientations can be derived by solving the eigenvector/eigenvalue problem of the matrix.

   The accuracy of the ellipsoid attribute calculation depends on the integration procedure. The accuracy of the integration depends on the number of nodes within a feature; the average position and variation in position is more accurate when we integrate over a large number of nodes. Thus, the accuracy of the attributes will also depend on the (local) grid density.

2. *Robustness of the extraction method with noise.* The presence of noise in data will introduce false positives, and false negatives in the collection of selected nodes. Besides an error in the attributes, this will cause the emergence of spurious features. The latter can be eliminated by choosing the right extraction parameters. The extraction parameters consist of the selection threshold value, the cluster threshold, and the connected component definition.

   - The selection threshold value (or multiple values) decides whether the data in a grid point satisfies our selection criterion. It can be set above the noise level in order to eliminate noise effects, but this will also influence the resulting feature.

   - The cluster threshold is the minimum number of nodes of a cluster; all clusters smaller than the cluster threshold are discarded. Thus, only large features remain, and small features resulting from noise are removed. However, we may also remove small but genuine features.

   - The connected component definition can be defined as: 1D-connected (where a node has 6 neighbours), 2D-connected (18 neighbours), and 3D-connected (26 neighbours). This definition is crucial in the clustering stage, since it determines if two adjacent nodes are in the same cluster or not. Obviously, 1D-connected will result in more and smaller clusters than 2D- or 3D-connected.

The extraction parameters must be chosen with care, therefore we will establish a number of guidelines for finding the right settings.

3 Synthetic data

In order to examine the relations between accuracy, noise, and extraction parameters, we created well-defined synthetic data on which we perform a number of experiments. The data is generated on a regular grid with a variable density. A scalar field is created on this grid with a variable initial value (set to zero by default), and noise is possibly added on top of this. The noise has a Gaussian distribution function with zero mean, and a variable standard deviation (SD); it is generated with an algorithm given by Press et al [5]. Furthermore, data values are added to grid nodes inside the synthetic features. The synthetic features are ellipsoids with given center position, axis lengths and orientations, plus a data value is defined at the center of the ellipsoid (set to 100.0 by default). The data within the ellipsoid decreases linearly from the center to the surface (value = 0.0). Thus, for each grid node inside the feature a data value is calculated and added to the present node value.

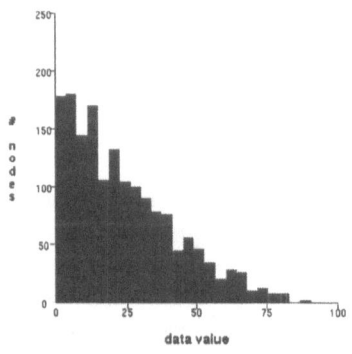

Fig. 3. Histograms of the generated synthetic ellipsoid data.

The synthetic data is used as the input in the feature extraction pipeline, where the data is thresholded, the selection clustered, and an ellipsoid-fit is performed around the clusters. Since the features in this data are predefined, the obtained attributes can be compared directly to the attributes specified as input, thus obtaining an experimental error estimation.

Figure 3 shows the histogram of the data within an ellipsoid feature (background with data = 0 is omitted from the histogram). Most of the nodes in the feature have a value close to zero and only few come close to the maximum value (=100). Additional noise will mostly affect the feature-nodes near the surface of the ellipsoid where the data values are small. Still it is possible to extract the feature since the maximum data value of the feature is significantly higher than the noise data. This is shown in figure 4, where a selection is made of nodes with a data value > 2*SD = 30.0, the figure shows the selected nodes by small crossmarks, and ellipsoids are fitted around each cluster with more than one node. One of the ellipsoids is significantly larger than the rest, this is the synthetic feature, it can be filtered out by choosing a larger cluster threshold, thus eliminating all small clusters.

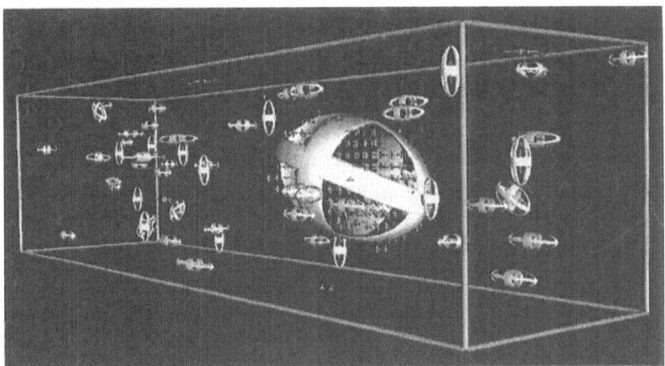

Fig. 4. Resulting selections from data with noise.

4 Experiments

4.1 Accuracy of the ellipsoid-fitting method

– **Center position.** The position detection is expected to have an error below
cell-size level, which can be proved by the following experiment. Synthetic
data is generated with a spherical feature with fixed radius and a position
moving in 50 steps from a corner node of a cell diagonal through the cell to
the center of the cell. Each of the 50 data sets are analysed by the feature ex-
tractor, and the resulting positions are used for error estimation. We expect
the resulting position to move stepwise through the cell, the steps are caused
by nodes entering or exiting the moving sphere. The distance to the diagonal
(the real position) divided by the diagonal length, gives a relative error for
the position detection. The same experiment can be repeated for different
grid resolutions, i.e. a feature with a larger number of selected nodes. This
will probably show that the accuracy is better for higher resolutions.
– **Axis length.** To determine the accuracy of the axis lengths, synthetic data
is generated containing ellipsoids with fixed orientation and with the radius
of one of the axes varying in one direction. Again, the variation is limited
within a cell, and the experiment is repeated for several grid resolutions.
Errors are calculated relatively to the cell size.
– **Axis orientation.** Synthetic data is generated containing ellipsoids with
fixed axes ratios with an eccentricity of 3:1:1, and with varying orientation
of the main axis (from 0 to 45 degrees), for several grid resolutions. Errors
are calculated relatively to the maximum possible angle, i.e. 45 degrees.

4.2 Robustness of the method

As discussed in section 2, there are three important settings in the feature extraction procedure, the selection threshold, the cluster threshold and the connected
component definition. The following experiments establish the relationships between these parameters, and the effects on the extracted features.

- **The selection threshold value.** Noise may introduce additional undesired clusters if the threshold value is set too low. The next experiment surveys the number of clusters found, as a function of the threshold value, and of the noise level. Synthetic data is generated with one feature, and for a number of different noise levels. Using this data, the number of clusters is monitored while slowly increasing the threshold until only one cluster (the synthetic feature) is found. The lowest threshold value that results in one feature is called the cut-off threshold value. It is an important value since it gives us the minimum threshold value that distinguishes the feature from the noise. The cut-off should be as low as possible, as higher threshold values result in smaller features. Therefore, the cut-off threshold will be used in further experiments, because it depends not only on the noise level, but also on the other feature extraction settings.
- **The cluster threshold.** The cluster threshold is a very useful parameter, since small irrelevant clusters are removed by it. In many cases (especially if noise is involved) the selection results in single unconnected nodes that just happen to satisfy the selection criteria, but are not significant. The cluster threshold is often an adequate remedy to filter out these undesired features. Therefore, we determine the cut-off threshold for different noise levels, as a function of the cluster threshold.
- **Connected component definition.** The neighbour definition will affect the number of clusters found. The 1D-definition is more strict than the others, and will result in more and smaller clusters, which amplifies the effects of the cluster threshold. In order to test this, the cut-off threshold is determined for all three definitions as a function of the cluster threshold, for one given noise level.

5 Results

5.1 The accuracy of the ellipsoid-fitting method

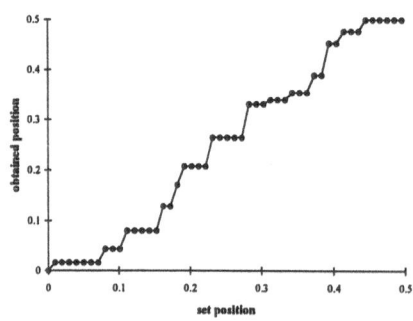

Fig. 5. Stepwise movement of the position within a cell.

First, the accuracy of the ellipsoid fitting method is established using the experiment described in section 4.1. Figure 5 shows the results of the accuracy tests for the position detection. The center position of the sphere starts at one corner node of the cell (relative position = 0), and ends in the center of the cell (relative position = 0.5). The obtained position is plotted as a function of the input position. It changes discontinuously every time a node enters or exits the moving sphere. Thus, a stepwise update of the position is found. The average distance to the diagonal is the average relative error of the position detection. Similar stepwise results are obtained for the axis and orientation detection. Since the attributes were varied within cell size, we may conclude that the ellipsoid-fit method detects shifts within sub-cell level.

As may be expected, the accuracy becomes better when the grid is more dense. Figure 6 shows the errors as a function of the number of selected nodes in the cluster. The figure clearly shows the exponential decrease of the error with respect to the number of nodes. The errors are below 7% when the clusters consist of more than 15 nodes. Thus, the ellipsoid attributes are accurate if a cluster threshold of 15 nodes is used.

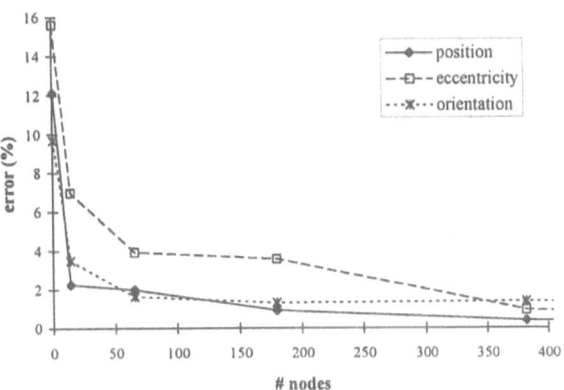

Fig. 6. Obtained errors for the ellipsoid attributes.

5.2 Robustness of the method

Now that the accuracy has been assured, the robustness of the method with respect to noise is investigated using the experiments described in section 4.2. First the number of clusters is determined as a function of the selection threshold value, and as a function of the noise level.

Figure 7 shows that for small thresholds a large number of clusters is found. This is an obvious results since the noise causes many node values to rise above the threshold level. If noise is added with an SD > 10, the number of clusters first increases with increasing threshold values because many nodes connect to form a large cluster which breaks up while increasing the threshold. In the end, a threshold value is found where only one feature remains. This value is the cut-off threshold; it becomes larger as the noise level increases.

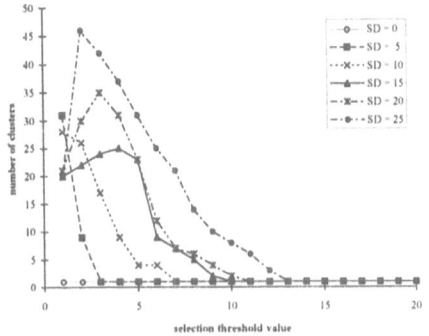

Fig. 7. The number of clusters as a function of the selection threshold value and the noise level.

The cut-off threshold is examined as a function of the noise level and the cluster threshold. Figure 8 shows that the cut-off threshold increases with increasing noise, still the cut-off threshold remains low for a large cluster threshold. Using a cluster threshold of at least 20 nodes, it suffices to use a selection threshold value of 1*SD in order to eliminate all clusters due to noise. If smaller features are expected, then a cluster threshold of 5 nodes in combination with a selection

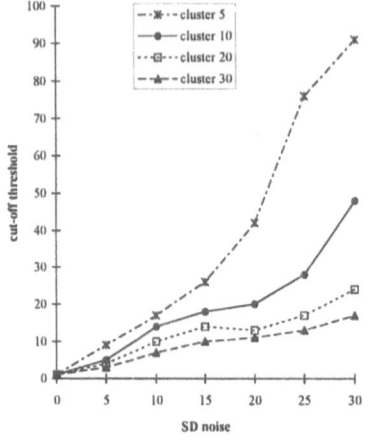 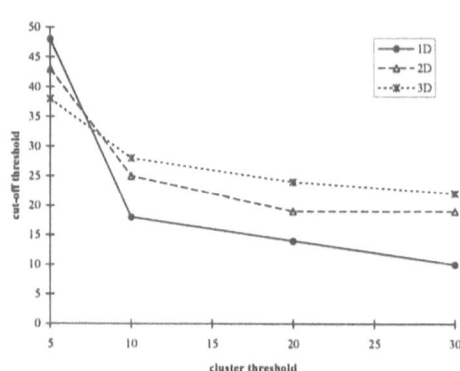

Fig. 8. The cut-off threshold value as a function of the noise level and the cluster threshold.

Fig. 9. The cut-off threshold value as a function of the connected component definition and the cluster threshold.

threshold value of at least 2*SD suffices, provided that the threshold value is significantly smaller than the maximum data value in the feature.

The next experiment determines the cut-off threshold for the three connected component definitions, as a function of the cluster threshold, with the noise level set to SD = 15.0. The result is plotted in figure 9, which shows that the cut-off threshold drops as the cluster threshold increases, and also that it drops faster if the connected component definition is set to 1D-connected. This definition results in more, smaller clusters which are easier to discard by the cluster threshold. Therefore, in case of noisy data, one should use the 1D-connected component definition.

5.3 Robustness of the calculated attributes

Finally, the effect of noise on the obtained ellipsoid attributes are investigated with optimal extraction parameters. Noise with an SD = 15.0 is added to the data, and 1D-connected component definition is used in combination with a cluster threshold of 15 nodes. In figure 10 errors are plotted as a function of the selection threshold value.

The figure clearly shows a large error in position for low selection threshold values. This is caused by the large cluster throughout the entire domain due to noise. Also, the error increases for large thresholds, caused by the fact that the feature is small and additional selected nodes due to noise affect the position significantly. Between the two extremes the errors are stable, thus the results of the method are relatively invariant to noise.

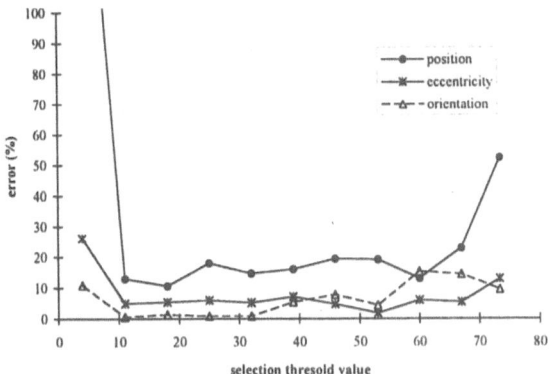

Fig. 10. The errors of the ellipsoid detection of data with noise (SD = 15.0).

6   Conclusions

During the execution of the experiments it became clear that a lot can be learned about the behaviour of the feature extraction method. Therefore, we consider this type of experiment extremely important for the exploration and validation of visualization techniques, and we recommend to do similar experiments with any new visualization method. In this case, the following conclusions can be drawn:

1. The ellipsoid attributes can be estimated with an accuracy below the cell-size level. The errors decrease for increasing grid density, i.e. for clusters with more nodes. A cluster threshold of 15 nodes results in errors below 7%.
2. In case of noisy data, the feature extraction parameters can be set in such a way that spurious features can be filtered out. A statictical analyses is needed in order to give the exact requirements, but based on these experiments the following guidelines for the feature extraction parameters can be given:
   - The cluster threshold is a powerful parameter to discard spurious features due to noise. Large cluster thresholds result in correct feature extraction, even close to the noise level. This is caused by the coherence in space of the selected nodes.
   - In case of noise, the use of the 1D-connected component definition is recommended, since this results in smaller clusters which are easier discarded by the cluster threshold.
3. The obtained ellipsoid attributes are stable despite the presence of noise. This means that the ellipsoid attributes are relatively invariant to noise.

7   Future research

The results of the experiments described in this paper pave the way for a number of interesting studies in the future.

- Small spurious features may be filtered out by morphological operators like opening and closing. This may enhance the effects of the cluster threshold.
- Further statistical analysis can be done on the extraction of features below noise level. Besides coherence in space, coherency in time may be exploited: e.g. if a feature is detected at one time, a prediction can be made of the feature some time later, this prediction can be used to extract the new feature. This suggests a predictive approach for feature tracking in time-dependent data.

Acknowledgments

This work is supported by the Netherlands Computer Science Research Foundation (SION), with financial support of the Netherlands Organization for Scientific Research.

References

1. D.C. Banks and B.A. Singer. A predictor-corrector technique for visualizing unsteady flow. *IEEE Trans. on Visualization and Computer Graphics*, 1(2):151–163, June 1995.
2. J.L. Helman and L. Hesselink. Visualization vector field topology in fluid flows. *IEEE Computer Graphics and Applications*, 11(3):36–46, 1991.
3. H.G. Pagendarm and B. Seitz. An algorithm for detection and visualization of discontinuities in scientific data fields applied to flow data with shock waves. In P. Palamidese, editor, *Scientific Visualization: Advanced Software Techniques*, pages 161–177. Ellis Horwood Limited, 1993.
4. F.J. Post, T. van Walsum, F.H. Post, and D. Silver. Iconic techniques for feature visualization. In G.M. Nielson and D. Silver, editors, *Proc. Visualization '95*, pages 288–295. IEEE Computer Society Press, 1995.
5. W.H. Press, S.A. Teukolsky, W.T. Vetterling, and B.P. Flannery. *Numerical Recipes in C: The Art of Scientific Computing*. Cambridge University Press, second edition, 1992.
6. F. Reinders, F.H. Post, and H.J.W. Spoelder. Feature extraction from pioneer venus ocpp data. In W. Lefer and M. Grave, editors, *Visualization in Scientific Computing '97*, pages 85–94. Springer Verlag, April 1997.
7. D. Silver and N.J. Zabusky. Quantifying visualizations for reduced modeling in nonlinear science: Extracting structures from data sets. *J. of Visual Communication and Image Presentation*, 4(1):46–61, March 1993.
8. T. van Walsum, F.H. Post, D. Silver, and F.J. Post. Feature extraction and iconic visualization. *Trans. on Visualization and Computer Graphics*, 2(2):111–119, 1996.

Enhancing the Visualization of Characteristic Structures in Dynamical Systems*

Helwig Löffelmann and Eduard Gröller**

Institute of Computer Graphics, Vienna University of Technology,
Karlsplatz 13/186/2, A-1040 Wien, Austria,
http://www.cg.tuwien.ac.at/home/

Abstract. We present a thread of streamlets as a new technique to visualize dynamical systems in three-dimensional space. A trade-off is made between solely visualizing a mathematical abstraction through lower-dimensional manifolds, i.e., characteristic structures such as fixed points, separatrices, etc., and directly encoding the flow through stream lines or stream surfaces. Bundles of streamlets are selectively placed near characteristic trajectories. An over-population of phase space with occlusion problems as a consequence is omitted. On the other hand, information loss is minimized since characteristic structures of the flow are still illustrated in the visualization.

Keywords: visualization, dynamical systems.

1   Introduction

Visualization [14] has become an established field of science during the past years. Dynamical systems, for example, flow fields, are an important topic concerning research in this area [2, 16]. Dynamical systems provide a mathematical framework to deal with the dynamics of a set of variables. They are used to model real world phenomena such as, e.g., the stock market, chemical reactions, or food chains.

A *dynamical system* is usually given by a vector of state variables which change over time [3]. If the formulas which describe the dynamics of the system are varying over time, a dynamical system is called *time-dependent*. If the rules guiding the dynamics are static over time, the dynamical system is called *steady* (time-independent). Usually a *continuous* dynamical system (also called *flow*) is specified by a set of ordinary differential equations (ODEs – $\dot{x} = f(x, p, t)$) together with a set of parameters (p). Often continuous dynamical systems are visualized in *phase space*, which is defined by associating each of the n state variables to one axis of an n-dimensional Cartesian coordinate system. In this

---

* http://www.cg.tuwien.ac.at/research/vis/dynsys/KnitDS97/
** mailto:helwig@cg.tuwien.ac.at, mailto:groeller@cg.tuwien.ac.at

paper we will concentrate on 3D continuous dynamical systems which are steady, i.e., function f does not depend on time t.

Several approaches to the visualization of dynamical system can be distinguished [11]. One class of techniques deals with the visualization of *characteristic elements* such as, e.g., fixed points, cycles, or separatrices. A structure of lower-dimensional objects is composed in phase space to describe the key features of the system's behavior [1]. For example, a separatrix is visualized to indicate two subsets of phase space with qualitatively different dynamics. A brief overview of the relation between local linearization and characteristic structures can be found in the Appendix.

Another class of approaches deals with the *direct visualization* of the system behavior. Integral curves visualize the evolution of specific initial settings which change according to the dynamics of the underlying flow. Many techniques are already available for the 2D case. Spot noise [18] and line integral convolution (LIC) [5], for example, provide an overview of 2D dynamics within a 2D domain. In 3D, however, direct visualization is difficult. Rendered images tend to be overloaded when entire portions of flow in 3D space are simultaneously visualized. Some attempts into this direction are illuminated stream lines [19] and volume-rendered 3D flow [7].

In addition to the visualization of characteristic elements and direct visualization, a third class of techniques deals with the representation of local properties [12]. Glyphs [6] represent certain quantities derived from the Jacobian matrix (local linearization of the flow) such as, e.g., acceleration, rotation, or divergence. Another approach [17] transforms a polygon positioned perpendicular to a trajectory to represent local information.

In this paper we present a technique which to a certain extent belongs to all of the three classes mentioned above. It was inspired by the concept of modeling knit-wear as yarn with a complex micro-structure [8]. We visualize the vicinity of characteristic trajectories, for example, the stream lines emanating from fixed points. A great number of short integral curves (streamlets) is used to directly code the system's behavior near the characteristic trajectory. By this approach of selectively placing streamlets we omit distracting image cluttering while still providing direct cues to the (local) system behavior. Visualizing the vicinity of characteristic stream lines enhances the abstract representation of the system's behavior by local cues of direct visualization.

2   A thread of streamlets

To come up with a useful technique of locally enhanced stream lines, we propose a model for the generation of a *thread of streamlets*. Near a predescribed stream line $\mathcal{T}$ (the *base trajectory*) many short streamlets are placed. Thereby a continuous representation of the system's behavior in the vicinity of the base trajectory is approximated.

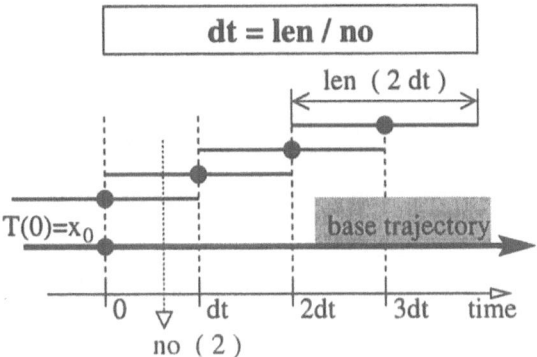

Fig. 1. Relation between streamlet density (no), streamlet integration length (len), and streamlet instantiation interval (dt)

Using constant flow as a reference model – stream lines are straight lines in this case – the thread of streamlets $\{T_i\}_{i\in\mathbb{N}}$ is defined as follows: Any cross-section perpendicular to base trajectory T is pierced by a constant number (no) of streamlets. Using integration time t as parameterization of base trajectory T ($T(0) = x_0$ = seed point of T), streamlets T_i are instantiated at time $t_i = i \cdot dt$ and integrated over the time interval $[i \cdot dt \pm \frac{len}{2}]$. See Fig. 1 for an illustration of the relationship between no, dt, and len, i.e., $dt = len/no$. Seed points $T_i(i \cdot dt)$ of newly instantiated streamlets are randomly chosen within a perpendicular cross-section through $T(i \cdot dt)$ corresponding to a probability distribution function (PDF) $d(\alpha, r)$ (see Eq. 1 and Fig. 2). In other words,

- many streamlets are arranged around a certain base trajectory T in a circular fashion. Thus, polar coordinates (r and α) were used to describe the seed states of the streamlets.
- Through PDF d the generated streamlet distribution is uniform within a certain radius (qR) and fades out linearly outside radius qR. This way of instantiating streamlets emphasizes the flow near base trajectory T.

$$d(\alpha, r) = \begin{cases} D & \text{if } 0 < r \leq qR \\ \frac{R-r}{R-qR}D & \text{if } qR < r \leq R \\ 0 & \text{if } R < r \end{cases} \qquad (1)$$

PDF $d(\alpha, r)$ is defined by parameters R (the maximal distance between $T(i \cdot dt)$ and $T_i(i \cdot dt)$) and $q \in [0, 1)$. The latter parameter is used to define PDF d as a truncated cone. This shape provides the fade-out characteristic of the streamlet placement procedure with respect to the distance from T. To guarantee that d is a PDF $\int d(\alpha, r) \, d\alpha \, dr$ must equal 1, i.e., the volume of the truncated cone must be 1. This constraint can be expressed as specification for parameter D:

$$D = \frac{3}{(1 + q + q^2)R^2\pi}$$

62

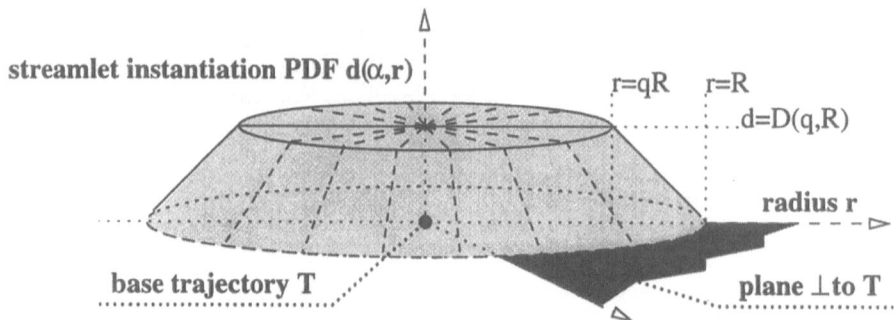

Fig. 2. Probability density function $d(\alpha, r)$ for the instantiation of streamlets based on a perpendicular cross-section through the base trajectory.

Computing a thread of streamlets for the reference model ($\dot{x} = $ **const.**), a bunch of line segments (streamlets – $\{\mathcal{T}_i\}_{i \in \mathbb{N}}$) of equal length ($len \cdot |\dot{x}|$) is generated. It this case of constant flow the streamlets are parallel to the base trajectory which is a straight line itself. The initial positions of streamlets $\{\mathcal{T}_i(i \cdot dt)\}_{i \in \mathbb{N}}$ are determined according to the PDF $d(\alpha, r)$. For any time t the cross-section perpendicular through $\mathcal{T}(t)$ is pierced by exactly $no = len/dt$ streamlets.

Applying this model to real (non-constant) flow data, local flow characteristics are visualized through the following variations from the constant flow reference setup:

- the **shape of the streamlets** directly visualizes the flow locally to the base trajectory. Local convergence/divergence or rotational behavior with respect to the base trajectory is intuitively depicted. Since local variations are significant in the area of (partial) degeneracies of the flow, characteristic trajectories are especially well suited to be chosen as base trajectories.
- the **streamlet length** is a direct visualization of flow velocities near the base trajectory. Due to this, the flow velocity can be depicted very well. Compared to color coding which is often used for velocity visualization the use of streamlets is more effective.

Taking a linear node repellor, i.e., a linear source, with eigenvalues 1, 10, and 100, for example, the flow characteristics in the vicinity of this fixed point can be visualized in different ways (see Fig. 3). Using threads of streamlets for a visualization of the characteristic trajectories – those which are aligned with the eigenvectors of the fixed point's Jacobian matrix – a dense and intuitive representation of the 3D flow near the fixed point is generated. Through the threads of streamlets (Fig. 3b) the flow next to the characteristic trajectories is visualized. A purely abstract notation (Fig. 3a) encodes the eigenvectors of the Jacobian matrix and the magnitudes of the associated eigenvalues. No information about the vicinity of the characteristic trajectories is provided.

3 Rendering

Drawing 1D objects poses several problems in the rendering stage. Shading, for example, improves the visual cues concerning the spatial arrangement of objects, but shading is usually defined on the basis of a surface (normal). Lines and curves have an infinite number of normals in each of their points. Therefore typical models such as Phong shading [15] can not be applied directly to 1D objects in 3D.

In 1989 Kajiya presented an "ad hoc" approach to deal with the problem of line shading in 3D which is based on an integration of all reflected intensities [9]. In 1996 Zöckler et al. described an efficient computation scheme for line shading in 3D which generates comparable results to the technique proposed by Kajiya [19]. A general framework for the task of shading k-dimensional manifolds in n-dimensional space was worked out by Banks in 1994 [4]. In addition to a consistent framework for the shading problem with arbitrary codimensions Banks also dealt with the problem of excess brightness-compensation which becomes an important topic if manifolds with codimension higher than 1 are shaded.

Another problem associated with line shading in 3D is (self-)shadowing. Normally, if shading 2D manifolds in 3D space, we (implicitly) deal with this aspect by assuming all surface points in (self-)shadow, where the outward normal $\mathbf{n}$ points away from the light vector $\mathbf{l}$, i.e., $\mathbf{n} \cdot \mathbf{l} < 0$. Furthermore we (implicitly) consider shadow rays before we compute surface shading. Both aspects are difficult with line shading in 3D. One approach to deal with these aspects comes from volume rendering: lines populating certain regions of 3D space can

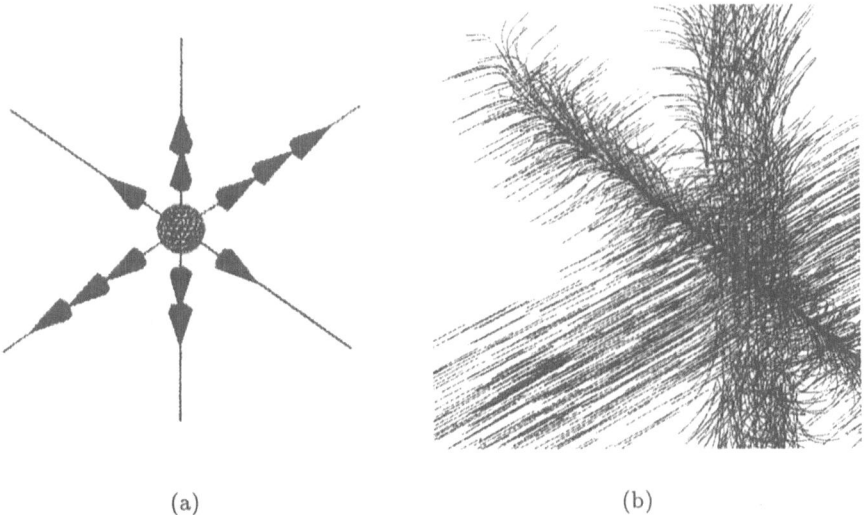

(a)   (b)

Fig. 3. Visualizing the flow near a linear node repellor in 3D: eigenvectors and eigenvalues (1, 10, and 100) (a), characteristic trajectories plus threads of streamlets (b).

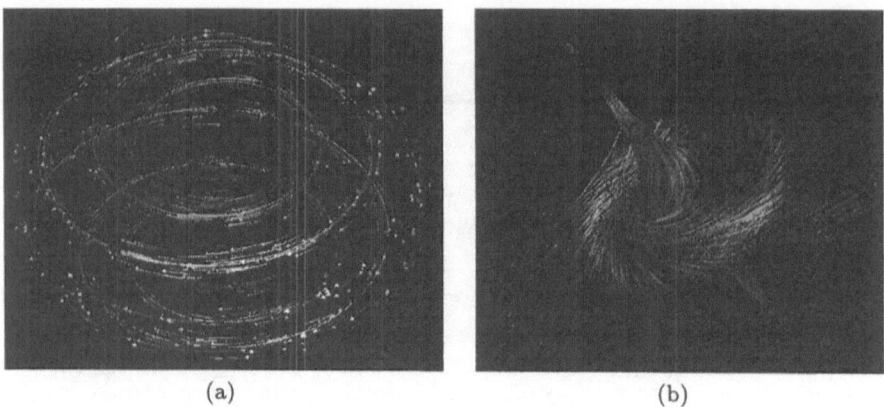

Fig. 4. A thread of streamlets visualizing the flow near a torus in 3D space (a); flow near a 3D focus visualized using two threads of streamlets (b).

be considered as volume opacity of a certain density. This assumption yields an exponential brightness attenuation for light passing through such a region. A paper by Max in 1995 compiles a comprehensive list of diverse models dealing with this effect [13].

⁻ For our implementation we chose the shading model used by Zöckler for shading the streamlets. Additionally we used depth cueing as a rough approximation of shadowing to enhance the spatial perceptability of the streamlets in 3D space. See Fig. 4a for an example. The heads of the streamlets have been pointed out by small arrow-heads to indicate the orientation of the flow. Furthermore color has been used to encode the flow velocity (blue ↔ slow, red ↔ fast). Line shading and depth cueing has been applied as described above.

4   Results

To test the newly proposed technique we firstly applied it to a simple cases, i.e., the fixed point of a linear dynamical system. Depending on the Jacobian matrix evaluated at this point, different results are obtained. Fig. 3b, for example, shows six threads of streamlets applied to the characteristic trajectories emanating from the fixed point. In this case the eigenvalues of the Jacobian matrix at the fixed point are 1, 10, and 100. The new visualization technique allows to easily depict the slow, medium, and fast directions of flow. Moreover, an impression is conveyed, how system states are repelled from the plane defined by the slow and medium direction (eigenvalues 1 and 10). Within that plane states are repelled from the slow direction which itself is therefore extremely instable in this setup. These flow characteristics typical for a dynamical system near a fixed point cannot be communicated by either showing an abstraction only (Fig. 3a) or a complete set of stream lines.

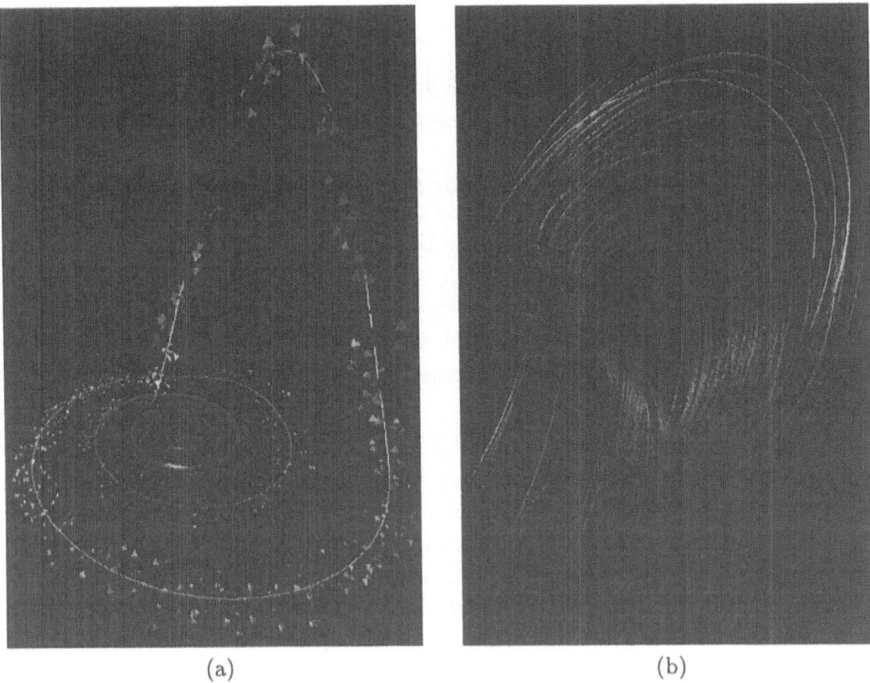

(a)   (b)

Fig. 5. Visualizing the flow velocity near a stream line of the Roessler system (a); visualizing the dynamics of a periodic dynamical system exhibiting a twisted torus (b).

Fig. 4b is generated by using two threads of streamlets for the visualization of a 3D focus, also within a linear dynamical system. The Jacobian matrix of this system exhibits one negative eigenvalue and two conjugate complex eigenvalues with positive real parts. System states are attracted along an instable 1D manifold – a line in the case of a linear system – and repelled into a stable 2D manifold (plane) perpendicular to the instable set. Applying the threads to both instable trajectories the dynamics near this fixed point are meaningful visualized. As in Fig. 4a color was used to encode flow velocity.

There is no need for applying the new technique only to characteristic trajectories. Fig. 5 shows two examples where different results were produced with this technique. The left image shows a thread of streamlets through the Roessler system. Instead of the streamlets themselves just arrow-heads at the end of each streamlets are displayed. Using size and color according to the velocity of the flow slow and fast areas within this system are intuitively visualized. The right image depicts the dynamics of a periodic flow near a twisted torus. Color coding indicates the velocity along the streamlets. As in Fig. 5a and 5b no characteristic trajectories were used, the evolution of the streamlets is more or less aligned with the base trajectory. Regions of local convergence/divergence are implicitly shown as areas with more/less streamlets.

5 Implementation

The technique presented in this paper was implemented within DynSys3D, a visualization system concerned with analytically specified dynamical systems in 3D space [10]. According to the modular concept of this system the new visualization technique is independent of the dynamical system and the numerical integrator specification. An AVS module is generated by linking the implementation of a specific dynamical system – basically two evaluation functions for calculating the flow vector and the Jacobian matrix at a specific system state – and a specific numerical integrator, for example, a Runge-Kutta integration scheme, to the thread of streamlets implementation. The module generates one thread of streamlets for a specific dynamical system by using a specific numerical integrator.

Parameters for the module are the starting location of the base trajectory $(\mathcal{T}(0))$ and its length (either temporal or spatial), the number of streamlets per cross-section (no), their length (len), the maximum distance of their seed-points (R) together with the fade-out parameter (q). The performance of this technique is between interactive and moderate (up to one or two minutes), depending on how many steamlets are computed. However, if parameter no temporarily is set to some small number, the visualization can be adjusted interactively.

6 Conclusions

We present a new technique for the visualization of dynamical systems, namely the use of a thread of streamlets for characteristic trajectories. This is useful, since a trade-off is made between only displaying structural information such as, e.g., fixed points and separatrices, and directly visualizing the system dynamics by the use of stream lines or stream surfaces. Since an abstract denotation of the dynamics caused by a dynamical system are very hard to understand for most users, enhancing this information by locally adding cues of direct visualization helps to communicate the crucial aspects of the system behavior.

Contrary to surface based stream line visualization techniques like the stream tube of sweep base trajectory representations threads of streamlets visualize the flow continuously in the vicinity of a stream line. Furthermore, using a thread of streamlets instead of entirely populating 3D phase space with stream lines, has the advantage of reducing occlusion. Although quite a number of papers deal with densely visualizing flow in 3D space, it seems to be necessary to place visual cues selectively to reduce occlusion problems. For high-quality versions of the images presented in this paper please visit the web page at URL http://www.cg.tuwien.ac.at/research/vis/dynsys/KnitDS97/.

Acknowledgements. The authors thank Markus Götzinger and Helmut Doleisch for their help in preparing this paper.

References

1. R. H. Abraham and C. D. Shaw. *Dynamics – The Geometry of Behavior*. Addison-Wesley, 2nd edition, 1992.
2. H. Aref, R. D. Charles, and T. T. Elvins. Scientific visualization of fluid flow. In C. A. Pickover and S. K. Tewksbury, editors, *Frontiers of Scientific Visualization*, pages 7–43. Wiley Interscience, 1993.
3. D. K. Arrowsmith and C. A. Place. *An Introduction to Dynamical Systems*. Cambridge University Press, 1990.
4. D. C. Banks. Illumination in diverse codimensions. *Computer Graphics*, 28(Annual Conference Series):327–334, 1994.
5. B. Cabral and L. Leedom. Imaging vector fields using line integral convolution. *Computer Graphics*, 27(Annual Conference Series):263–270, 1993.
6. W. C. de Leeuw and J. J. van Wijk. A probe for local flow field visualization. In *Proceedings of IEEE Visualization '93*, pages 117–123, October 1994.
7. T. Frühauf. Raycasting vector fields. In *Proceedings of IEEE Visualization '96*, pages 115–120, 1996.
8. E. Gröller, R. T. Rau, and W. Straßer. Modeling and visualization of knitwear. *IEEE Transactions on Visualization and Computer Graphics*, 1(4):302–310, December 1995.
9. J. T. Kajiya and T. L. Kay. Rendering fur with three dimensional textures. *Computer Graphics*, 23(Annual Conference Series):271–280, July 1989.
10. H. Löffelmann and E. Gröller. DynSys3D: A workbench for developing advanced visualization techniques in the field of three-dimensional dynamical systems. In *Proceedings of The Fifth International Conference in Central Europe on Computer Graphics and Visualization '97*, pages 301–310, Plzen, Czech Republic, February 1997.
11. H. Löffelmann, E. Gröller, R. Wegenkittl, and W. Purgathofer. Classifying the visualization of analytically specified dynamical systems. *Machine GRAPHICS & VISION*, 5(4):533–550, 1996.
12. H. Löffelmann, Z. Szalavári, and E. Gröller. Local analysis of dynamical systems – concepts and interpretation. In *Proceedings of The Fourth International Conference in Central Europe on Computer Graphics and Visualization '96*, pages 170–180, Plzen, Czech Republic, February 1996.
13. N. Max. Optical models for direct volume rendering. *IEEE Transactions on Visualization and Computer Graphics*, 1(2):99–108, June 1995.
14. G. M. Nielson and B. Shriver. *Visualization in Scientific Computing*. IEEE Computer Society Press, 1990.
15. B.-T. Phong. Illumination for computer generated pictures. *CACM June 1975*, 18(6):311–317, 1975.
16. F. H. Post and T. van Walsum. Fluid flow visualization. In H. Hagen, H. Müller, and G. M. Nielson, editors, *Focus on Scientific Visualization*, pages 1–40. Springer, 1993.
17. W. J. Schröder, C. R. Volpe, and W. E. Lorensen. The stream polygon: A technique for 3D vector field visualization. In *Proceedings of IEEE Visualization '91*, pages 126–132, October 1991.
18. J. J. van Wijk. Spot noise – texture synthesis for data visualization. *Computer Graphics*, 25(Annual Conference Series):309–318, July 1991.
19. M. Zöckler, D. Stalling, and H.-C. Hege. Interactive visualization of 3D-vector fields using illuminated streamlines. In *Proceedings of IEEE Visualization '96*, pages 107–113, October 1996.

Appendix: Fixed Points and Characteristic Trajectories

Assuming $\dot{\mathbf{x}} = \mathbf{f_p}(\mathbf{x})$ to be continuous and steady dynamical system in 3D space, the fixed points $\mathbf{o}_i$ of $\mathbf{f_p}$ are given by

$$\dot{\mathbf{o}}_i = \mathbf{f_p}(\mathbf{o}_i) = 0$$

Using the Taylor expansion of $\mathbf{f_p}$ in the vicinity of a fixed point $\mathbf{o}_i$ together with local linearization a linear ODE in terms of $\Delta = \mathbf{x} - \mathbf{o}_i$ can be derived:

$$\mathbf{f_p}(\mathbf{o}_i + \Delta) = \sum_{k=0}^{\infty} \frac{1}{k!} (\Delta \cdot \nabla)^k * \mathbf{f_p}\big|_{\mathbf{o}_i} \approx \underbrace{\mathbf{f_p}(\mathbf{o}_i)}_{=0} + \nabla \mathbf{f_p}\big|_{\mathbf{o}_i} \cdot \Delta$$

$$\underbrace{\dot{\mathbf{o}}_i}_{=0} + \dot{\Delta} = \frac{d(\mathbf{o}_i + \Delta)}{dt} = \mathbf{f_p}(\mathbf{o}_i + \Delta)$$

$$\implies \qquad \dot{\Delta} = \nabla \mathbf{f_p}\big|_{\mathbf{o}_i} \cdot \Delta$$

This linear dynamical system can be investigated by analyzing $\mathbf{f_p}$'s Jacobian matrix $\nabla \mathbf{f_p}\big|_{\mathbf{o}_i}$ at the fixed point $\mathbf{o}_i$. One possibility is to determine the eigenvalues and eigenvectors of the Jacobian. They completely describe the dynamics of a linear dynamical system [1].

Transferring the results from local linearization to the original system, we facilitate the fact that (in the hyperbolic case) manifolds spanned by the eigenvectors of $\nabla \mathbf{f_p}\big|_{\mathbf{o}_i}$ are coplanar with $\mathbf{f_p}$'s characteristic manifolds through $\mathbf{o}_i$. Characteristic stream lines, for example, are trajectories which are attracted to a saddle fixed point $\mathbf{o}_i$ while all the other stream lines near the characteristic trajectory (finally) are repelled from $\mathbf{o}_i$.

Editor's Note: see Appendix, p. 144f. for colored figures of this paper

Part III
Flow Visualization

Particle Tracing in σ-Transformed Grids using Tetrahedral 6-Decomposition

I. Ari Sadarjoen[1], Alex J. de Boer[1,2], Frits H. Post[1], and Arthur E. Mynett[2]

[1] Dept. of Computer Science, Delft University of Technology
PO Box 356, 2600 AJ Delft, The Netherlands
email: {I.A.Sadarjoen, A.J.deBoer, F.H.Post}@cs.tudelft.nl
[2] WL | Delft Hydraulics, Strategic R&D
PO Box 177, 2600 MH Delft, The Netherlands
email: Arthur.Mynett@wldelft.nl

Abstract. *Particle tracing in curvilinear grids often employs decomposition of hexahedral cells into 5 tetrahedra. This method has shortcomings when applied to σ-transformed grids, a grid type having strongly sheared cells, commonly used in hydrodynamic simulations. This paper describes an improved decomposition method into 6 tetrahedra. It is shown that this method is robust in σ-transformed grids, however large the shearing. Results are presented of applying the technique to a real world simulation. Comparisons are made between the accuracy and speed of the 5-decomposition and the 6-decomposition methods.*

1 Introduction

Particle tracing is an important technique for visualization of velocity vector fields resulting from computational fluid dynamics (CFD) simulations [3]. The basic particle tracing technique is based on a stepwise numerical integration of the velocity field. Especially the numerical integration techniques have been well studied [7,8]. A source of complications is the use of irregular grids in CFD simulations, such as structured curvilinear grids. The cells of such grids are hexahedra with arbitrarily deformed geometry; the adjacency structure (topology) of the grid cells is regular. But the irregular geometry causes critical operations in particle tracing algorithms to be more complex, such as finding which cell contains a given point.

Several solutions have been proposed for this problem. One is to transform the deformed hexahedral cells to cubes (computational space or C-space), perform the path integration in a regular grid, and transform the results back to the original deformed grid. Another solution is to perform the tracing directly in the deformed grid (physical space or P-space), using a decomposition of the hexahedral cells into tetrahedra. We have analysed and compared these techniques in an earlier paper [6], and found that the P-space approach gave the best results both in accuracy and efficiency, if visualization is applied as a postprocessing activity following the actual computations, and C-space velocities are not directly available.

The P-space algorithms using a tetrahedral decomposition of hexahedral cells have gained acceptance [4], but some problems can occur in strongly deformed grid cells. In large-scale hydrodynamic simulations, the x- and y-dimensions are typically 2 to 3 orders of magnitude larger than the z-dimension. Such simulations often use so-called σ-transformed curvilinear grids, in which the hexahedral cells can be very flat and strongly deformed. This can cause the tetrahedral 5-decomposition to produce locally invalid results, and the P-space tracing algorithm to fail.

In this paper, we will examine the reasons for this failure in more detail, and propose a new decomposition which can be proven to maintain validity of the grid even for very strong deformations of σ-transformed grid cells. Section 2 gives some fundamentals of P-space particle tracing, and briefly surveys the 5-tetrahedron decomposition of hexahedral cells. After explaining the σ-transformation in Section 3, we analyse the problems occurring with the 5-decomposition. Section 4 presents the improved 6-decomposition, and experimental results are shown in Section 5. Conclusions and directions for further work are given in Section 6.

2 Fundamentals of Particle Tracing

The computation of a particle path is based on the integration of the ordinary differential equation

$$\frac{d\mathbf{x}}{dt} = \mathbf{v}(\mathbf{x}) \tag{1}$$

where $\mathbf{x}$ denotes the position of the particle, t is time, and $\mathbf{v}(\mathbf{x})$ the velocity field. The starting position $\mathbf{x}_0$ of the particle provides the initial condition at initial time t_0: $\mathbf{x}(t_0) = \mathbf{x}_0$. Subsequent points are calculated as

$$\mathbf{x}(t_{n+1}) = \mathbf{x}(t_n) + \int_{t_n}^{t_{n+1}} \mathbf{v}(\mathbf{x})dt \tag{2}$$

using a numerical integration method. The solution is a sequence of particle positions $(\mathbf{x}(t_0), \mathbf{x}(t_1), \ldots)$ at time steps $t_0, t_1, \ldots$

Particle Tracing in Regular Rectangular Grids

Let us first recall the structure of a particle tracing algorithm for regular rectangular (uniform) grids:

> find cell containing initial position   (*point location*)
> **while** particle in grid
>   determine velocity at current position   (*interpolation*)
>   calculate new position   (*integration*)
>   find cell containing new position   (*point location*)
> **endwhile**

Point location is the process of determining which cell contains a specified point. In uniform grids, this is easily accomplished by splitting the coordinates of a point into the integer cell indices (i, j, k) and fractional offsets (α, β, γ). *Interpolation* is the process of determining a data value at an arbitrary position in a given cell, using the surrounding grid nodes and the fractional offsets. In uniform grids, this is typically done with first-order trilinear interpolation.

Particle Tracing in Curvilinear Grids

In practice, many CFD applications do not use uniform grids, but *structured curvilinear grids*, consisting of deformed, hexahedral cells. An advantage of curvilinear grids is that they can follow the shape of curved or complex geometries such as airplane wings and coast lines. The disadvantage is that algorithms working on these grids are more complex, because the cells are no longer regular cubes, but they may be sheared and have curved faces.

One strategy often applied in many CFD simulation systems, is to transform the curvilinear grids in physical space to a uniform grid in a new domain, called *computational space*. Unfortunately, for visualization algorithms, this method did not turn out to be beneficial, as was investigated in detail in [6].

Another strategy is to calculate the particle path directly in physical space. This would avoid transformations between the two domains, although at the expense of more difficult point location and interpolation. Interpolation in curvilinear grids is more difficult, because the offsets are harder to determine in a curved cell. Point location in curvilinear grids is more difficult, because there is no longer a direct relation between the global coordinates of a point and the cell indices. Instead, a search must be performed, by checking for several cells if they contain the point. Usually, there is a previous position in a known cell, which is connected to the new position in the unknown cell by a line. Along this line, the algorithm traverses subsequent adjacent cells by intersecting the line with the cell faces and checking which adjacent cell has that face in common.

Decomposition into 5 Tetrahedra

One way to cope with curved cells works by decomposing the hexahedral cells into tetrahedra. The advantages of tetrahedra is that they are convex and planar, which facilitates containment tests and face intersection tests.

The simplest and most efficient scheme is to decompose the hexahedral cells into 5 tetrahedra, henceforth called the *5-decomposition*. Figure 1a shows a cube which is decomposed into 1 (shaded) center tetrahedron and 4 corner tetrahedra. In a structured grid, the decomposition can be done in two directions. To ensure connection of cell faces and to avoid overlapping cells, these two directions should be alternated in adjacent cells, as shown in Figure 1b.

In tetrahedra, interpolation and point location are performed as follows. *Interpolation* in tetrahedra is done using linear interpolation. Figure 1c shows a tetrahedron ABCD, where α, β, γ denote the fractional offsets in the tetrahedron, with the restriction that $\alpha + \beta + \gamma \leq 1$. If $\mathbf{v}_A$ is the data value in node A, $\mathbf{v}_B$

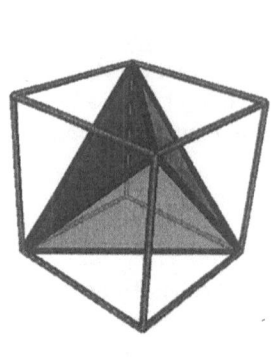 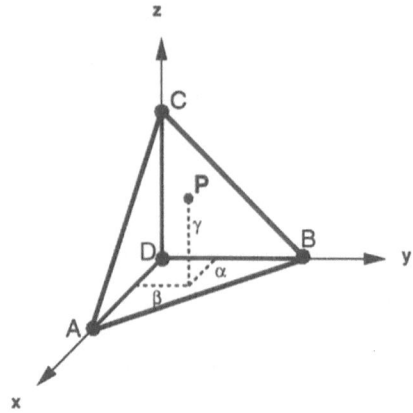

(a) 1 center tetrahedron and 4 corner tetrahedra

(b) Linear interpolation in tetrahedra

Fig. 1. Decomposition of a hexahedral cell into 5 tetrahedra

the data value in node B, etc., then the interpolated value v_P in some position P in the tetrahedron is: $v_P = v_D + \alpha(v_A - v_D) + \beta(v_B - v_D) + \gamma(v_C - v_D)$. The fractional offsets (α, β, γ) may be found by inverting the interpolation of the position of P in the tetrahedron.

$$P = D + \alpha(A - D) + \beta(B - D)\gamma(C - D) \tag{3}$$

$$(\alpha, \beta, \gamma) = (A - D|B - D|C - D)^{-1}(P - D) \tag{4}$$

Point location is again done by traversing cells from a previous position, although not entire cells are traversed, but the tetrahedra into which they are decomposed.

3 σ-Transformed Grids

Point location using tetrahedral 5-decomposition regularly fails, especially in a specific type of grids known as σ-*transformed grids*. In our test cases, up to 40%(!) of the particles were caught in an infinite loop between two cells, or stopped completely. Before explaining the cause of these problems, let us first describe this type of grids.

σ-Transformed grids are frequently used in hydrodynamic simulations of shallow waters, such as marine coasts or estuaries. They consist of stacked 2D xy-layers, each of which is a well-formed quadrangular mesh with curved and usually orthogonal grid lines. Corresponding nodes in different layers have identical x,y coordinates. In the vertical direction, the grid lines are straight and parallel to the z-axis. σ-coordinates are defined relative to the local water elevation ζ and depth

d, as $\sigma = \frac{z-\zeta}{\zeta+d}$. The top grid layer, where $\sigma = 0$, follows the free water surface, which usually only varies gradually. The bottom layer, where $\sigma = -1$, follows the sea floor geometry, which typically has strongly varying depths throughout the model. The layers in between are constructed with a prescribed distribution. Figure 2 shows one possible distribution of 6 layers. Figure 3 shows a sea floor geometry and a vertical grid slice in the Lith harbour data set, which was used in a simulation and visualization project at WL | Delft Hydraulics [5] (see Figure 9 for colour (see Appendix)).

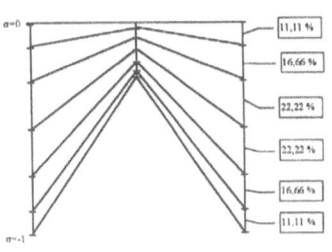

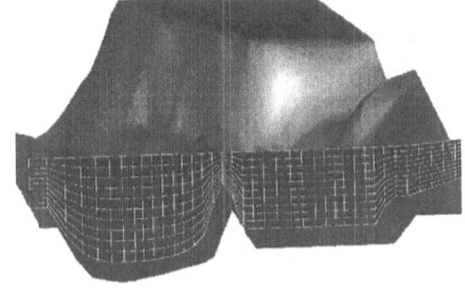

Fig. 2. Distribution of a σ-transformed grid with 6 layers

Fig. 3. σ-transformed grid, with a sea floor geometry and a vertical grid slice

In σ-transformed grids, many cells are sheared in the vertical direction, because the number of layers is constant, while the local depth varies, so parallel vertical edges often lie at very different depths. The shearing is increased as the cells are typically very thin in these applications: the model may be hundreds of kilometers wide and only tens of meters deep. Strongly sheared cells have some typical characteristics. In a normal cell, the orientation of the center tetrahedron is as shown in Figure 4a, but in a strongly sheared cell, the center tetrahedron has been turned inside out, as shown in Figure 4b. The top faces BEG and DEG now lie at the bottom, while the bottom faces BDE and BDG lie on top. The edges BD and GE have crossed each other. This is possible because the center tetrahedron has edges spanning the entire cell.

These strongly sheared cells result in two problems with the 5-decomposition. One problem is that the center tetrahedron overlaps with corner tetrahedra, and even with a neighbouring cell. As a consequence, the point location algorithm cannot determine a unique tetrahedron which contains a given point, and exits. The second problem is that particles may get caught in an infinite loop between two tetrahedra. Due to the reversed orientation of the center tetrahedron, the point location algorithm fails to find the correct outgoing face, and therefore the correct adjacent tetrahedron. As a consequence, the algorithm moves from a corner tetrahedron to the center tetrahedron, and then return to the corner tetrahedron where it came from, instead of proceeding to the next one. In this way, it will continue moving back and forth forever.

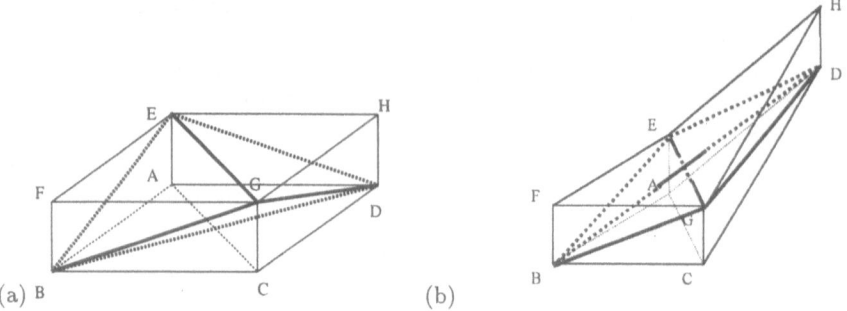

Fig. 4. (a) Normal (a) and (b) reversed tetrahedral orientation in 5-decomposed hexahedra

The frequency at which these problems occurred, varied between 4% and 40% of the particles, depending on the data set and particle source locations. In some cells, the problems might be solved by changing the decomposition direction, since the problem is direction dependent. But then the problem would probably occur in other cells, because the direction is chosen globally for the entire grid.

4   Tetrahedral 6-Decomposition

An apparent solution to the point location problems comes to mind: scale or shear a deformed cell such that the twisted orientation of the edges and tetrahedra is avoided. However, scaling or shearing grid cells amounts to applying a computational space algorithm: the grid is transformed to a different domain, where the cells are regular and rectangular. We chose not to do this in Section 2 because of the loss of accuracy and efficiency [6].

A better approach is to use a different tetrahedral decomposition. A systematic overview of the possibilities can be found in [1]. A hexahedron can be decomposed into 5, 6, and any even number between 12 and 24 tetrahedra. For reasons of efficiency and storage space, the preferred approach is the decomposition into 6 tetrahedra, henceforth called the *6-decomposition*. Figure 5 shows how this is accomplished: a hexahedral cell is decomposed into two three-sided prisms, each of which is decomposed into 3 tetrahedra. Just like the 5-decomposition, the 6-decomposition has two directions: each face diagonal can be chosen in two ways. But an advantage of the 6-decomposition method is that it does not require the directions to alternate for adjacent cells.

The main advantage of this 6-decomposition method is that it solves the point location problems. There is no longer a center tetrahedron whose edges span the entire cell, and which may cross each other when the cell is sheared in the vertical direction. Figure 6 shows the 6-decomposition in a normal cell, and in a sheared cell comparable to Figure 4b. It can be clearly seen that the tetrahedra in the

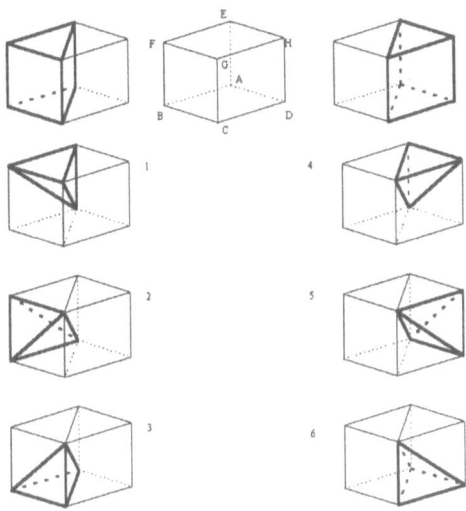

Fig. 5. Decomposition of a hexahedron into 6 tetrahedra

sheared prism retain their orientations, since the hatched planes AGH and ADG retain their orientations and relative positions. No tetrahedron has been turned inside out. It can be shown that this method is robust: the tetrahedra will never change orientation, no matter how large the shearing is, as long as the edges are only displaced in the vertical direction (as is the case with σ-transformed grids).

5 Results

The technique described above was implemented in a set of AVS/Express modules called PLANKTON-97 [2]. Modules were developed for interactively placing point, line, and plane particle sources, for calculating the particle paths, and for creating animations. To evaluate the technique, three types of tests were performed: a functional test and a speed test.

Functional Test

To test the system, we have performed tests in artificial and real world data sets. Whereas the 5-decomposition method would fail in 4% to 40% of the released particles, the 6-decomposition method did not have any problems in tracing the particles through strongly deformed cells. Here, we show an example of particles traced in the Lith Harbour data set. The grid is a σ-transformed curvilinear grid with 121x40x10 cells. At the grid nodes, velocity and turbulence intensity were defined. In this data set, 100 particles were released in a horizontal plane. Figure 7 shows the particles rendered as arrows. It is clearly visible that

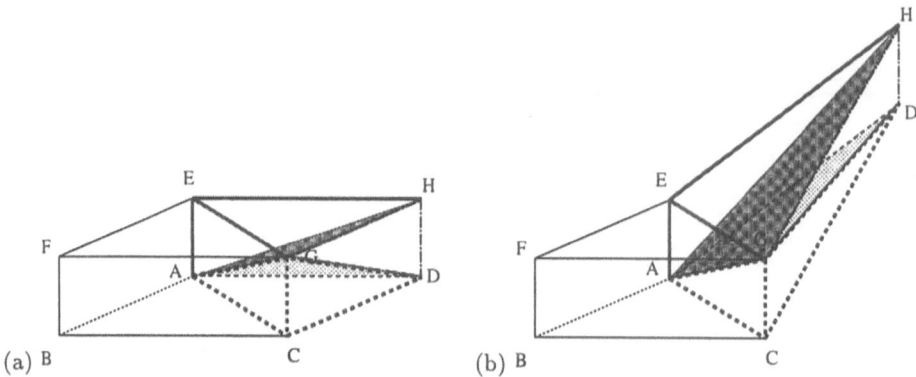

Fig. 6. Tetrahedral orientation in 6-decomposed (a) normal cell and (b) sheared cell

the 5-decomposition method failed to trace 14 out of 100 particles, rendered as circles. The 6-decomposition was successful for all particles (not shown).

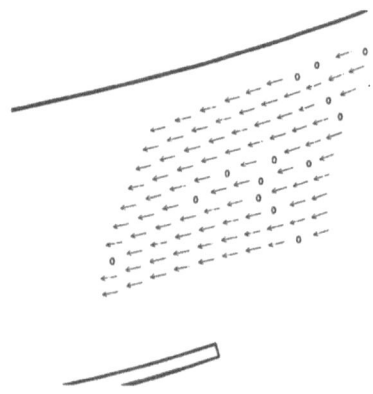

Fig. 7. The 5-decomposition method failed to trace 14% of the particles in Lith Harbour

Figure 8 shows another example where the 6-decomposition was successful (see Figure 10 for colour (see Appendix)). Here, instead of the particles themselves, the traced particle paths were rendered, coloured with velocity magnitude. In addition, a bounding box and a sea floor geometry were rendered to increase the sense of depth.

Speed Test

To compare the speed of both algorithms, we measured the execution times necessary for creating 100 animation frames. In the Lith Harbour data set, particles were released from one source located near the center of the data set.

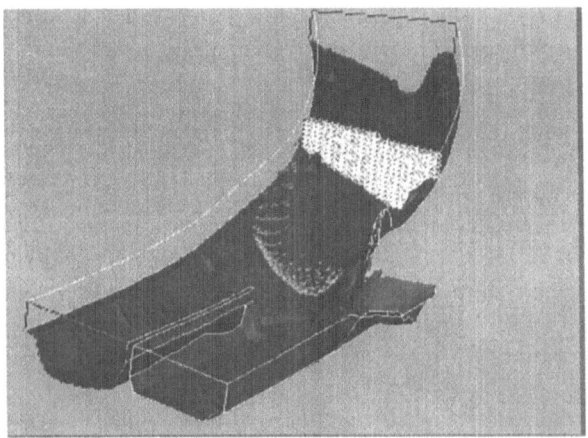

Fig. 8. Particles successfully traced with the 6-decomposition method

After every 25 frames, a new particle was released. For each frame, we calculated 25 integration steps of $\Delta t = 10s$, which amounts to 2500 integration steps. The machine used was an SGI Indigo[2] with a MIPS R10000 processor at 195 MHz. Note that the 6-decomposition is slightly faster than the 5-decomposition, even though it creates more tetrahedra. This is due to a simple optimization for avoiding redundant operations in traversing the decomposed cell. Table 1 lists the test results, which were performed several times and averaged, to obtain accurate measurements.

	5-decomposition	6-decomposition
execution time (s)	10.61	10.47
# traversed cells	85	85
# traversed tetrahedra	234	243

Table 1. Speed comparison of 5-decomposition and 6-decomposition methods

6 Conclusions and Future Work

It has been shown that decomposition of hexahedral cells in σ-transformed grids into 6 tetrahedra is better than decomposition into 5 tetrahedra. Particles whose paths could not be traced due to the limitations of the 5-decomposition, could be successfully traced with the 6-decomposition method. The 6-decomposition method has shown to be robust, regardless of the amount of vertical shearing of the cells.

In practice, the grids used in hydrodynamic simulations have more application-specific features, such as missing (dry) grid points, thin dams, boundary points requiring special care, etc. However, these do not fall within the scope of this paper, but some solutions are presented in Chapter 6 of [2].

The tetrahedral decomposition can be used in unstructured tetrahedral grids with only slight modifications, if face/cell adjacency information is available for traversing the grid.

Acknowledgements

The work described here is a part of the second author's Master's thesis work [2], performed at WL | Delft Hydraulics. We wish to thank Irving Elshoff of WL | Delft Hydraulics for his supervision and valuable suggestions.

References

1. G. Albertelli and R.A. Crawfis. Efficient subdivision of finite-element datasets into consistent tetrahedra. In R. Yagel and H. Hagen, editors, *Proc. Visualization '97*, pages 213–219. IEEE Computer Society Press, 1997.
2. A.J. de Boer. Reconstructie en uitbreiding van Plankton in AVS/Express. Master's thesis, Delft University of Technology, January 1998. In Dutch.
3. A.J.S. Hin and F.H. Post. Visualization of turbulent flow with particles. In G.M. Nielson and R.D. Bergeron, editors, *Proceedings Visualization '93*, pages 46–52. IEEE Computer Society Press, Los Alamitos, CA, 1993.
4. D.N. Kenwright and D.A. Lane. Optimization of time-dependent particle tracing using tetrahedral decomposition. In G.M. Nielson and D. Silver, editors, *Proc. Visualization '95*, pages 321–328. IEEE Computer Society Press, 1995.
5. D.G. Meijer. Lock approach second ship lock at Lith. Scale model investigation and DELFT3D-calculcations. Technical report, WL | Delft Hydraulics, June 1995. In Dutch.
6. A. Sadarjoen, T. van Walsum, A.J.S. Hin, and F.H. Post. Particle tracing algorithms for 3D curvilinear grids. In *Proc. 5th EuroGraphics Workshop on Visualization in Scientific Computing*, 1994.
7. T. Strid, A. Rizzi, and J. Oppelstrup. Development and use of some flow visualization algorithms. In *Computer Graphics and Flow Visualization in Computational Fluid Dynamics*, Lecture Series 1989-07. Von Kármán Institute for Fluid Dynamics, 1989.
8. C. Teitzel, R. Grosso, and T. Ertl. Efficient and reliable integration methods for particle tracing in unsteady flows on discrete meshes. In W. Lefer and M. Grave, editors, *Visualization in Scientific Computing '97*, pages 31–41. Eurographics, Springer, 1997.

Editor's Note: see Appendix, p. 146 for colored figures of this paper

Particle Tracing on Sparse Grids

Christian Teitzel, Roberto Grosso, and Thomas Ertl

Computer Graphics Group, University of Erlangen
Am Weichselgarten 9, 91058 Erlangen, Germany

Abstract. These days sparse grids are of increasing interest in numerical simulations. Based upon hierarchical tensor product bases, the sparse grid approach is a very efficient one improving the ratio of invested storage and computing time to the achieved accuracy for many problems in the area of numerical solution of differential equations, for instance in numerical fluid mechanics. The particle tracing algorithms that are available so far cannot cope with sparse grids. Now we present an approach that directly works on sparse grids. As a second aspect in this paper, we suggest to use sparse grids as a data compression method in order to visualize huge data sets even on small workstations. Because the size of data sets used in numerical simulations is still growing, this feature makes it possible that workstations can continue to handle these data sets.

1 Introduction

In 1990 sparse grids were introduced by Zenger [10]. With their help it is possible to reduce the total amount of data points or the number of unknowns in discrete partial differential equations. Due to these benefits, sparse grids are more and more used in numerical simulations nowadays [1–4].

On the other hand, it is rather difficult to visualize the results of the simulation process directly on sparse grids, since evaluation and interpolation of function values is quite complicated on such grids. Because of this, up to the present the results of numerical simulations on sparse grids are extrapolated to the associated full grid. Then, all known visualization algorithms on full grids can be performed, e.g. particle tracing, iso-surface extraction, volume rendering, etc.. However, a major drawback of this procedure is the fact that the advantage of low memory consumption of sparse grids comes to nothing using the associated full grid for the visualization step.

Therefore, visualization tools working directly on sparse grids are going to be an important topic of research. Heußer and Rumpf already started working on iso-surface extraction on sparse grids [7]. The first aim of our work is to introduce particle tracing directly on sparse grids (Section 3). Furthermore, a second aspect of this work is the idea that sparse grids can be used for data compression in order to visualize huge data sets on small workstations (Section 4). Additionally, the results of error, time, and memory analyses are listed in Section 4. In order to introduce particle tracing on sparse grids, new methods and classes had to be developed. This special class hierarchy is described in Subsection 3.1. In Subsection 3.2 we describe the implementation of our sparse grid classes as modules within the framework of the IRIS Explorer visualization environment.

2 Basics of Sparse Grids

In this section a brief summary of the basics of sparse grids is given. For a detailed survey of sparse grids we refer to [1, 10]. In order to make this overview easy to understand and to reduce the number of indices, we describe only three-dimensional grids, whereas the sketches reveal the one- and two-dimensional situations.

Let $f : [0,1]^3 \longrightarrow \mathbf{R}$ be a smooth function defined on the unit cube in $\mathbf{R}^3$ with values in $\mathbf{R}$. Furthermore, f should vanish on the boundary of the cube. This condition is not a strong restriction but is just helpful for an elegant description. Of course, our program can handle three-dimensional functions and even vector fields without zero boundary conditions. If such a function f is stored in the computer memory, then function values at certain positions on a spatial grid are stored in an array. The simplest mesh is a uniform one. Now let G_{i_1,i_2,i_3} be a uniform grid with respective mesh widths $h_{i_j} = 2^{-i_j}$, $j = 1,2,3$. On these grids we can introduce the following partial ordering relation: G_{i_1,i_2,i_3} is a refinement of G_{k_1,k_2,k_3} if and only if $k_j \leq i_j$, $j = 1,2,3$. and $k_1 + k_2 + k_3 < i_1 + i_2 + i_3$. Thus we obtain a hierarchy of meshes.

Now let $\hat{L}_n$ be the function space of the piecewise tri-linear functions defined on $G_{n,n,n}$ and vanishing on the boundary. Additionally, consider the subspaces S_{i_1,i_2,i_3} of $\hat{L}_n$ with $1 \leq i_j \leq n$, $j = 1,2,3$. which consist of the piecewise tri-linear functions defined on G_{i_1,i_2,i_3} and vanishing on the grid points of all coarser grids. Apparently, the hierarchy of grids naturally introduces a hierarchy of subspaces and it follows:

$$\hat{L}_n = \bigoplus_{i_1=1}^{n} \bigoplus_{i_2=1}^{n} \bigoplus_{i_3=1}^{n} S_{i_1,i_2,i_3} \quad .$$

Hence, we have found a hierarchical basis decomposition of the function space $\hat{L}_n$. Piecewise tri-linear finite elements are used as basis functions in each subspace S_{i_1,i_2,i_3}. We define the basis functions (Figure 1) of the subspace S_{i_1,i_2,i_3} of $\hat{L}_n$:

$$b_{k_1,k_2,k_3}^{(i_1,i_2,i_3)}(x_1,x_2,x_3) := \prod_{j=1}^{3} w_{i_j}(x_j - m_{k_j}^{(i_j)}) \quad \text{with} \quad m_{k_j}^{(i_j)} = (2k_j - 1) \cdot h_{i_j} \quad ,$$

$$1 \leq k_j \leq 2^{i_j - 1} \quad , \quad \text{and} \quad w_i(x) := \begin{cases} \frac{h_i + x}{h_i} & : \quad -h_i \leq x \leq 0 \\ \frac{h_i - x}{h_i} & : \quad 0 \leq x \leq h_i \\ 0 & : \quad \text{else} \end{cases} \quad .$$

Now we are interested in some estimations of the interpolation error. Hence, let

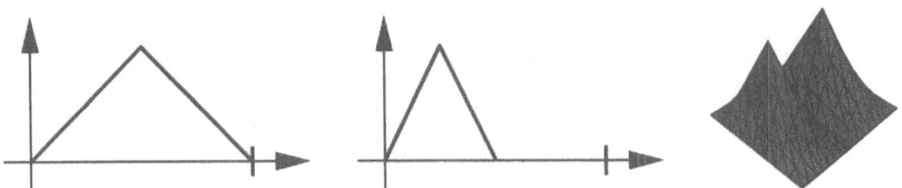

Fig. 1. Examples of basis functions, $b_1^{(1)}$ and $b_1^{(2)}$ on the left and $b_{1,1}^{(1,2)}$ and $b_{1,2}^{(1,2)}$ on the right hand side.

$\hat{f}_n \in \hat{L}_n$ be the interpolated function on the grid $G_{n,\ldots,n}$. Then, $\hat{f}_n$ is given by

$$\hat{f}_n = \sum_{i_1=1}^{n}\sum_{i_2=1}^{n}\sum_{i_3=1}^{n} f_{i_1,i_2,i_3} \quad \text{where} \quad f_{i_1,i_2,i_3} = \sum_{k_1=1}^{2^{i_1-1}}\sum_{k_2=1}^{2^{i_2-1}}\sum_{k_3=1}^{2^{i_3-1}} c_{k_1,k_2,k_3}^{(i_1,i_2,i_3)} \cdot b_{k_1,k_2,k_3}^{(i_1,i_2,i_3)}.$$

The values $c_{k_1,k_2,k_3}^{(i_1,i_2,i_3)}$ are called contribution coefficients and $f_{i_1,i_2,i_3} \in S_{i_1,i_2,i_3}$ is a linear combination of the basis functions of the appropriate subspace. It can be shown that the following estimations hold with regard to the L^2 and L^∞ norms (compare [1, pp. 13]):

$$\|f_{i_1,i_2,i_3}\|_2 \leq \frac{1}{27} \left\| \frac{\partial^6 f}{\partial x_1^2 \partial x_2^2 \partial x_3^2} \right\|_2 \cdot h_{i_1}^2 h_{i_2}^2 h_{i_3}^2 \quad , \tag{1}$$

$$\|f_{i_1,i_2,i_3}\|_\infty \leq \frac{1}{8} \left\| \frac{\partial^6 f}{\partial x_1^2 \partial x_2^2 \partial x_3^2} \right\|_\infty \cdot h_{i_1}^2 h_{i_2}^2 h_{i_3}^2 \quad , \tag{2}$$

$$\left\|f - \hat{f}_n\right\|_2 \leq \frac{1}{243} \left\| \frac{\partial^6 f}{\partial x_1^2 \partial x_2^2 \partial x_3^2} \right\|_2 \cdot h_n^2 = O\left(h_n^2\right) \quad , \tag{3}$$

$$\left\|f - \hat{f}_n\right\|_\infty \leq \frac{1}{72} \left\| \frac{\partial^6 f}{\partial x_1^2 \partial x_2^2 \partial x_3^2} \right\|_\infty \cdot h_n^2 = O\left(h_n^2\right) \quad . \tag{4}$$

So far we have just dealt with regular uniform meshes, which are named full grids. Now let us turn to sparse grids. Consider the subspaces S_{i_1,i_2,i_3} with $i_1 + i_2 + i_3 = const$. Equations (1) and (2) show that $\|f_{i_1,i_2,i_3}\|_\infty$ and $\|f_{i_1,i_2,i_3}\|_2$ have a contribution of the same order of magnitude, namely $O(2^{-2\cdot const})$ for all subspaces with $i_1 + i_2 + i_3 = const$. Additionally, these subspaces have the same number of basis functions, namely $2^{const-3}$. Since the number of basis functions is equivalent to the number of stored grid points and because of the contribution argument as well, it seems to be a good idea to define a sparse grid space $\tilde{L}_n$ as follows:

$$\tilde{L}_n := \bigoplus_{i_1+i_2+i_3 \leq n+2} S_{i_1,i_2,i_3}.$$

Now the interpolated function $\tilde{f}_n \in \tilde{L}_n$ is given by

$$\tilde{f}_n = \sum_{i_1+i_2+i_3 \leq n+2} f_{i_1,i_2,i_3} \tag{5}$$

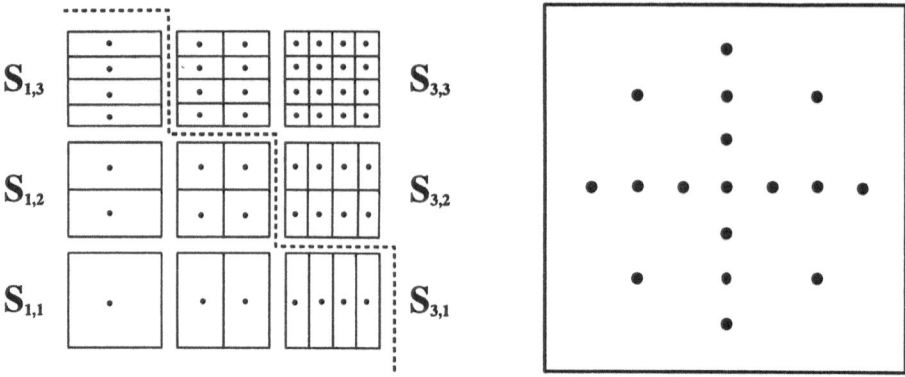

Fig. 2. On the left hand side a two-dimensional hierarchical subspace decomposition is shown and on the right hand side you can see the respective sparse grid.

and the interpolation errors with regard to the L^2 and L^∞ norms are given by (compare [1, pp. 23])

$$\left\| f - \tilde{f}_n \right\|_2 \le \left(1 + \frac{3}{4}n + \frac{9}{16}\binom{n+1}{2} \right) \frac{h_n^2}{9^3} \left\| \frac{\partial^6 f}{\partial x_1^2 \partial x_2^2 \partial x_3^2} \right\|_2 \tag{6}$$

$$= O\left(h_n^2 \left(\log_2 \left(h_n^{-1} \right) \right)^2 \right), \tag{7}$$

$$\left\| f - \tilde{f}_n \right\|_\infty \le \left(1 + \frac{3}{4}n + \frac{9}{16}\binom{n+1}{2} \right) \frac{h_n^2}{6^3} \left\| \frac{\partial^6 f}{\partial x_1^2 \partial x_2^2 \partial x_3^2} \right\|_\infty \tag{8}$$

$$= O\left(h_n^2 \left(\log_2 \left(h_n^{-1} \right) \right)^2 \right). \tag{9}$$

These estimations show that the sparse grid interpolated function $\tilde{f}_n$ is nearly as good as the full grid interpolated function $\hat{f}_n$.

Now we consider the dimensions of the function spaces $\hat{L}_n$ and $\tilde{L}_n$, which correspond to the number of nodes of the underlying grids. Obviously, the dimension of the full grid space is given by

$$dim\left(\hat{L}_n \right) = O\left(2^{3n} \right) = O\left(h_n^{-3} \right). \tag{10}$$

For the sparse grid the following equation holds:

$$dim\left(\tilde{L}_n \right) = O\left(2^n \cdot n^2 \right) = O\left(h_n^{-1} \left(\log_2 \left(h_n^{-1} \right) \right)^2 \right). \tag{11}$$

Therefore, a tremendous amount of memory is saved if sparse grids are used instead of full grids (see Section 4).

If the function f is given and a certain accuracy is required, then it is possible to use $\hat{f}_n \in \hat{L}_n$ or $\tilde{f}_m \in \tilde{L}_m$ where m is just slightly greater than n. Due to the very low memory consumption of sparse grids, it is better to use the function $\tilde{f}_m$. On the other hand the function f is often given in discrete form as data set on a full grid. In this case it is not possible to reach a better accuracy with the sparse grid approach than with the original full grid data. However, equations (7), (9), and (11) show that a very small loss of accuracy is rewarded with a huge amount of saved storage.

3 Particle Tracing on Sparse Grids

Flow visualization tools based upon particle methods continue to be an important utility of flow simulation. Additionally, the importance of sparse grids in numerical simulations is still growing. However, so far particle tracing algorithms could only handle data sets given on full grids. Now we present a particle tracer that can cope with sparse grids. Our new particle tracing module supplies the same features, e.g. colored streak lines, ribbons, tubes, balls, and tetrahedra (see Figure 3), as our previous full grid particle tracing tool, which is partially described in [5] and [6].

3.1 Class Hierarchy for Efficient Interpolation on Sparse Grids

Lagrange visualization techniques of a vector field v are based upon the numerical solution of an initial value problem for the differential equation: $dx/dt = v(x, t)$. Usually, a numerical integration method is used to obtain a solution. All such methods have in common that they must evaluate the vector field v at certain positions, which are in general not at grid points. Therefore, the value of v at such a position has to be interpolated. As mentioned in Section 2, this interpolation on sparse grids is different from that one on full grids, whereas the other parts of the particle tracing algorithm can remain unchanged.

In contrast to the tri-linear full grid interpolation, the sparse grid interpolation does not operate locally, because one basis function in every subspace contributes to the function value. Since the tri-linear interpolation is one of the most time consuming operations during the particle tracing process on full grids [8], the complicated sparse grid interpolation is all the more time consuming. Therefore, it is important to execute the interpolation as fast as possible.

Normally, the contribution coefficients of the sparse grid are stored in a binary tree [1, 2, 7]. Then, a recursive tree traversal has to be performed in order to interpolate the function value. This tree traversal is very slow. Although caching strategies can increase the efficiency of the traversal [7], the computation of the values remains rather time consuming.

Hence, the contribution values are not stored in a binary tree but in arrays. Then, it is not necessary to traverse a tree but the required contribution coefficient can be accessed directly. Therefore, we have implemented a particular C++ class hierarchy. Due to the limited amount of space, we can just give a very brief idea of the classes.

Initially, recall that the sparse grid space $\tilde{L}_n$ is the direct sum of all subspaces $S_{i,j,k}$ with $i + j + k \leq n + 2$. Now we define the *level of a subspace* as the number $n = i + j + k - 2$. Moreover, we define a *level of the sparse grid space* as the

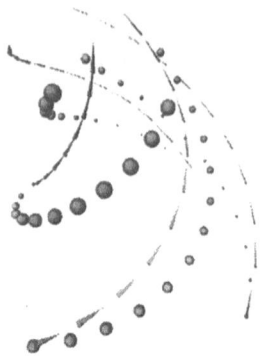

Fig. 4. Streak tubes in a cavity flow; the red tubes are computed on a full grid of level 7, the other tubes are created on sparse grids of level 7 (yellow), 5 (blue), and 3 (green).

Fig. 3. Colored streak balls and tetrahedra in a vortex flow given on a sparse grid.

direct sum of all subspaces of the same level of subspaces. Therefore, $\tilde{L}_n$ is the direct sum of its first n levels and is called a *sparse grid of level n*.

Besides abstract base classes, classes for input, and other auxiliary classes, the classes of interest are named hbSparseGrid, hbLevel, and hbSubspace. The class hbSparseGrid contains a stack of n levels of class hbLevel. Furthermore, hbLevel comprises the respective number of subspaces $((n + 1)n/2)$, denoted hbSubspace. The class hbSubspace contains an array of the size 2^{n-1} times data dimension, where the contribution coefficients are stored. The function value at an arbitrary position is computed by means of formula (5). In order to compute a function value, the class hbSparseGrid contains a method calcValue(...). This method sends a 'calcValue()' to each hbLevel to accumulate the contributions to the resulting value. Then, the method hbLevel::calcValue(...) performs a loop over all subspaces of the current level. In this loop, the required basis function is determined by means of the coordinates of the current position. Recall that only one basis function per subspace is unequal to zero at a certain position because all basis functions are hat-functions. Hence, we know the required contribution value. Now the 'height' over the current position in the tri-linear hat-function is determined and multiplied with the contribution value. Thus, we obtain the total contribution of this subspace to the function value. Additionally, we compute the Jacobian, which is needed to compute the local rotation of the flow for displaying bands and tetrahedra, in this loop by looking up the correct 'height' of the derivative of the hat-function, a simple box-function. The efficiency of this implementation is shown in Section 4.

3.2 Implementation as IRIS Explorer Module

Our new particle tracer, which works on data sets given on sparse grids, is implemented as an IRIS Explorer module and named StreakbandHB. As integration

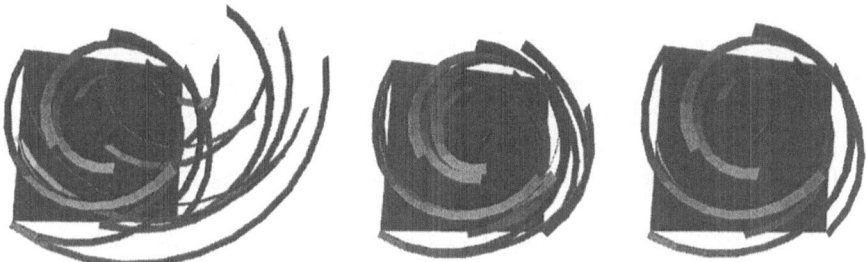

Fig. 5. Streak bands in a vortex flow; ribbons containing blue edges display the flow on a full grid of level 7, bands with green edges the flow on sparse grids of level 0 (left), 1 (middle), and 4 (right); the ribbons computed on full and sparse grids coincide on screen for levels greater than 3.

methods for the particle tracing algorithm of `StreakbandHB`, we use the integration schemes that we have already implemented in our full grid particle tracer, called `Streakband`. A comparison of these schemes can be found in [9]. An adaptive Runge-Kutta method of order 3 (RK3(2)) is used for the tests described in Section 4.

In order to visualize the particles, we have chosen the same geometrical primitives as in our full grid particle tracing module, namely lines, bands, tubes, balls, and tetrahedra. Of course, all kinds of traces can visualize an additional scalar value by means of color coding. Moreover, balls and tetrahedra can reveal another scalar value by their size. Besides that, bands and tetrahedra display the local vorticity of the flow via rotating around the actual streak line. Since both modules, `Streakband` and `StreakbandHB`, are provided with the same functionality, their results can be compared easily (see Section 4).

Besides the actual particle tracer, some additional modules had to be implemented in order to handle sparse grids properly. First of all, a module, called `DemoSparseGridHB`, is needed to create an analytical demo vector field on a sparse grid of a certain level. Secondly, a function, denoted `LatToSparseGridHB`, is used in order to transfer a full grid given as Explorer `cxLattice` data type to a sparse grid. Finally, `PrintSparseGridHB` is a helpful tool for debugging sparse grid routines.

In order to allow these new modules sending and receiving sparse grid data via the Explorer network, a new Explorer data type has been created, named `HBSparseGrid3D`.

4   Results

In order to compare our sparse grid particle tracing module with full grid particle tracers, two data sets were used. The first one, which was provided by S. H. Enger from the Lehrstuhl für Strömungsmechanik of the University of Erlangen, is a cavity flow data set on a full grid of level 7, i.e. 129^3 nodes (see Figure 4). The data set contains the velocity, pressure, and temperature at each vertex. Hence, it consumes more than 40 MB. Notice that the same data set with a resolution of 8 levels would need more than 320 MB, that is too much for most workstations. On the other hand, this data set stored on a sparse grid of level 7 consumes only 175 kB.

The second data set is an analytic one. It is a vortex flow (compare Figures 3 and 5). Since the data set is analytical, we are able to create sparse and full grids in any resolution only limited by the main memory of the used machine. Therefore, we chose the analytic vector field for our quantitative efficiency tests. Nevertheless, the performance of the compared modules was nearly the same while testing on the cavity data set.

All tests were performed on a Silicon Graphics computer with a 196 MHZ R10000 processor. For testing, at each time nine streak ribbons were computed consisting of about 500 particles (see Figure 5). The computing time of StreakbandHB is compared with that of our full grid Streakband module and of the NAG-Advect module, which is provided together with the IRIS Explorer. The CPU-times were measured in seconds and are listed in the following table.

level	2	3	4	5	6	7
points of full grid	5^3	9^3	17^3	33^3	65^3	129^3
StreakbandHB (sparse grid)	0.15 s	0.28 s	0.47 s	0.95 s	1.65 s	4.61 s
Streakband (full grid)	0.42 s	0.87 s	1.60 s	3.22 s	6.51 s	13.26 s
NAG-Advect (full grid)	1.09 s	1.33 s	1.61 s	1.89 s	2.28 s	2.66 s

Table 1. Computing times in CPU-seconds using an analytic vortex flow.

The used integration methods were an adaptive Runge-Kutta scheme RK3(2) in case of our Streakband modules and an adaptive Runge-Kutta scheme RK4(5) in case of the NAG-Advect program. See [9] for a discussion of different integration algorithms for particle tracing.

At first glance, it is astonishing that the full grid Streakband module is slower than our sparse grid StreakbandHB module. This is due to the fact that Streakband is adjusted to multi-block curvilinear grids. In order to cope with such grids, the stencil walk algorithm is performed during the particle tracing. This algorithm is unnecessary on uniform grids and therefore not performed by StreakbandHB. Thus, it is not fair to compare the computational times of those modules, but anyway the full grid Streakband is needed for the comparison of the actual particle traces.

The measured times show that interactive particle tracing is possible even on sparse grids of level 7. Secondly, the table reveals the drawback of sparse grid interpolation that the computing time exponentially rises if the level of the grid is increased. In contrast to this, the computing time of the NAG-Advect module is growing slowly. In theory, the time for particle tracing on full grids is independent of the grid size.

Now the accuracy of sparse grid particle tracing is considered. Therefore, the traces computed by StreakbandHB are compared with their counterparts resulting from Streakband. Recall that the error of full grid interpolation can be estimated at $O(h^2)$ and that of sparse grid interpolation at $O((h \cdot \log_2(h^{-1}))^2)$. This is a rather small difference. Moreover, the integration error of RK3(2) is of order $O(\tau^3)$ where τ denotes the current time step [9]. From this point of view, it does not seems to be too bad using sparse instead of full grid particle tracing. In fact, the results of particle tracing on the analytic data set confirm these estimations because the ribbons computed on full and sparse grids coincide on screen for levels greater than 3 (compare Figure 5).

However, during the deduction of the mentioned upper bounds of the interpolation errors, the smoothness of the data was needed (compare equations (3), (4), (6), and (8)). Since discrete data sets are not smooth at all, these estimations do not hold in case of discrete data. Indeed, Figure 4 reveals that the particle traces computed on sparse grids converge rather slowly to the full grid solution. Nevertheless, due to the great advantage of low memory consumption, it is possible to use a sparse grid of quite a high level to overcome this problem.

The great benefit of the sparse grid technique is the low number of required grid points. The next table shows the memory consumption of a typical data set resulting from a numerical flow simulation. Assume that five floating point values, namely three velocity components, pressure, and temperature, are given at each grid node. Then, these floating point values add up to 20 bytes per node. Thus, we obtain the following results:

level	5	6	7	8	9	10
points of full grid	33^3	65^3	129^3	257^3	513^3	1025^3
sparse grid	29 kB	73 kB	175 kB	415 kB	970 kB	2.2 MB
full grid	640 kB	5 MB	40 MB	320 MB	2.5 GB	20 GB

Table 2. Memory consumption of a typical data set.

This table shows that sparse grids are very suitable for compressing huge data sets. By dint of this, it is possible to visualize such data even on small workstations.

5 Conclusion

We have introduced particle tracing on sparse grids. This allows to carry out flow visualization directly on sparse grids without transforming the results of numerical simulations on sparse grids to the associated full grids. Secondly, the sparse grid approach can be used as a compression method in order to realize particle tracing in huge data sets on small workstations.

There are several directions of future work. The first aim is to introduce further visualization techniques on sparse grids. First of all, we are going to introduce volume rendering on these grids. A second goal is to enlarge the field of applications. At the moment, we are thinking about a particle tracing algorithm on curvilinear sparse grids. Furthermore, we intend to implement adaptive sparse grids with error monitoring. Last but not least, there are possibilities to accelerate the sparse grid interpolation by sophisticated caching strategies. On the one hand, a pre-computing mechanism of a certain number of levels could be implemented. On the other hand, pre-computing a certain number of cells could be advantageous.

References

1. H.-J. Bungartz. *Dünne Gitter und deren Anwendung bei der adaptiven Lösung der dreidimensionalen Poisson-Gleichung.* PhD thesis, TU Munich, 1992.
2. H.-J. Bungartz and T. Dornseifer. Sparse grids: Recent developments for elliptic partial differential equations. Technical report, TU Munich, 1997.
3. M. Griebel, W. Huber, U. Rüde, and T. Störtkuhl. The combination technique for parallel sparse-grid-preconditioning or -solution of pde's on multiprocessor machines and workstation networks. In L. Bougé, M. Cosnard, Y. Robert, and D. Trystram, editors, *Second Joint International Conference on Vector and Parallel Processing*, pages 217–228, Berlin, 1992. CONPAR/VAPP, Springer-Verlag.
4. M. Griebel, M. Schneider, and C. Zenger. A combination technique for the solution of sparse grid problems. In P. de Groen and R. Beauwens, editors, *International Symposium on Iterative Methods in Linear Algebra*, pages 263–281, Amsterdam, 1992. IMACS, Elsevier.
5. R. Grosso, M. Schulz, and T. Ertl. Fast and accurate visualization of steady and unsteady flows. Technical Report 3, University of Erlangen, 1996.
6. R. Grosso, M. Schulz, J. Kraheberger, and T. Ertl. Flow visualization for multiblock multigrid simulations. In *Virtual Environments and Scientific Visualization '96*, Heidelberg, 1996. Springer-Verlag.
7. N. Heußer and M. Rumpf. Efficient visualization of data on sparse grids. In H.-C. Hege and K. Polthier, editors, *Visualization and Mathematics*, Berlin. Springer-Verlag. In preparation.
8. D. N. Kenwright and D. A. Lane. Optimization of Time-Dependent Particle Tracing Using Tetrahedral decomposition. In G. M. Nielson and Silver D., editors, *Visualization '95*, pages 321–328, Los Alamitos, CA, 1995. IEEE Computer Society, IEEE Computer Society Press.
9. C. Teitzel, R. Grosso, and T. Ertl. Efficient and Reliable Integration Methods for Particle Tracing in Unsteady Flows on Discrete Meshes. In W. Lefer and M. Grave, editors, *Visualization in Scientific Computing '97*, pages 31–41, Wien, April 1997. Springer-Verlag. Proceedings of the Eurographics Workshop in Boulogne-sur-Mer, France.
10. C. Zenger. Sparse grids. In *Parallel Algorithms for Partial Differential Equations: Proceedings of the Sixth GAMM-Seminar*, Kiel, 1990.

Editor's Note: see Appendix, p. 147 for colored figures of this paper

Visualization of Time-Dependent Velocity Fields by Texture Transport

Joachim Becker[1] and Martin Rumpf[2]

[1] Institute for Applied Mathematics , Freiburg University , Germany
[2] Institute for Applied Mathematics , Bonn University , Germany

Abstract. Vector field visualization is an important topic in scientific visualization. The aim is to graphically represent field data in an intuitively understandable and precise way, which should be closely related to the physical interpretation. A new tool, the texture transport method is presented, which especially applies to time-dependent velocity fields. It is based on an accurate numerical scheme for convection equations, which is used to compute Lagrangian coordinates in space time. These coordinates are then used as texture coordinates referring to some prescribed texture in the Lagrangian reference space. The method is combined with a reliability indicator. This indicator influences the final appearance of the texture and thereby leads to reliable visual information. At first the method applies to 2D problems. It can be generalized to 3D.

1 Introduction

The visualization of field data, especially of velocity fields from CFD computations is one of the fundamental task in scientific visualization. A variety of different approaches has been presented. The simplest method to draw vector plots at nodes of some overlayed regular grid in general produces visual clutter, because of the typically different local scaling of the field in the spatial domain, which leads to disturbing multiple overlaps in certain regions, whereas in other areas small structures such as eddies can not be resolved adequately. The central goal is to obtain a denser, intuitively better receptable method. Furthermore it should be closely related to the mathematical meaning of field data, which is mainly expresses in its one to one relation to the corresponding flow. If a vector field $v : \Omega \times I\!\!R_0^+ \to I\!\!R^n$ for some domain $\Omega \subset I\!\!R^n$ is given, and for simplicity we at first assume that $v \cdot \nu = 0$ where ν is the outer normal on $\partial\Omega$ (lateron we will define corresponding in- and outflow conditions in a Lagrangian frame) then the corresponding flow $\phi : \Omega \times I\!\!R_0^+ \to \Omega$ is described by the system of ordinary differential equations

$$\partial_t \phi(x,t) = v(\phi(x,t),t)$$

and the initial condition $\phi(x,0) = x$.
The spot noise method proposed by van Wijk [22] introduces spot like texture splats which are aligned by deformation to the velocity field in 2D or on surfaces

in 3D. These splats are plotted in the fluid domain showing strong alignment patterns in the flow direction. The originally first order approximation to the flow was improved by de Leeuw and van Wijk in [6], where they use higher order polynomial deformation of the spots in areas of significant vorticity. The Line Integral Convolution (LIC) approach of Cabral and Leedom [2] integrates the above ODE forward and backward in time at every pixelized point in the domain, convolves a white noise along these particle paths with some Gaussian type filter kernel and takes the resulting value as an intensity value for the corresponding pixel. According to the strong correlation of this intensity along the stream-line and the lack of any correlation in the orthogonal direction the resulting texturing of the domain shows dense streamline filaments of varying intensity. Hege and Stalling [19] increased the performance of this method especially by reusing portions of the convolution integral already computed on points along the streamline. Max et al. [12] proposed a similar method on surfaces. Max and Becker [13] present a method for visualizing 2D and 3D flows by animating tex-tures. Turk [21] discusses an approach by which a certain number of streamlines is automaticly equally distributed on the computational domain.

Especially for 3D velocity fields particle tracing is a very popular tool. But a few particle integrations released by the user can hardly scope with the complex-ity of 3D vector fields. Zöckler et al. [20] use pseudo randomly distributed and illuminated and transparent streamlines to give a denser and receptable repre-sentation, which shows the overall structure and enhances important details.

An effective method to calculate stream surfaces in 3D, which nicely depicts sep-aration phenomena has been presented by Hultquist [9]. Van Wijk [23] proposed the implicit stream surface method. For a stationary flow field the transport equations $v \cdot \nabla \phi = 0$ are solved for given v and certain inflow and outflow boundary conditions in a precomputing step. Then isosurfaces of the resulting function ϕ are streamsurfaces and can efficiently be extracted with interactive frame rates even for larger data sets.

Most of these methods are designed and implemented on flow fields, which are constant in time. If we for instance apply line integral convolution in the time-dependent case successive images of a time sequence are in general not correlated. Grey level values at grid points change very rapidly because the streamlines at time t and $t + \delta t$ on which the convolution is performed have almost no overlap even for very small δt. Therefore we ask for an approach using texture based methods as well-suited tools to ensure a overall representation of field data, which avoids the above drawback in the non stationary case. We adopt the idea of the implicit streamsurfaces and discuss the corresponding transport problem for time-dependent data, solve it numerically for certain boundary and initial conditions and use the result to generate an appropriate texture mapping.

At the inlet of a fluid container we prescribe inflow boundary conditions, which are the inflow coordinates, respectively the inflow time. Furthermore outflow boundary conditions are given at the outlet and slip conditions on the remaining part of the boundary. In the interior the linear transport equations with respect to the prescribed velocity $v(x, t)$ describe the fluid motion, i.e. the transport of the inflow time and inflow coordinates. The set of points in space and time which shares a specific inflow coordinate coincides with a particle line, whereas the set of points with the same inflow time and inflow coordinates on a bounded surface respectively line on the inlet, describe the movement of the corresponding surface or line in time. Therefore in 2D we take the space spanned by the inflow time and the inflow coordinates as texture space and prescribe a texture with strong correlation in the direction of time. Then using the numerical results of the transport calculation, in explicit the numerical inflow time and inflow coordinates as texture coordinates we obtain a dense representation of particle lines in terms of visible texture correlation. This representation continuously depends on time and we can easily animate the evolution. In 3D we proceed similar as in the implicit streamsurface method proposed by van Wijk and texture the resulting streamsurfaces analogously.

The paper is organized as follows. In Section 2 we will in detail explain the continuous transport problem and the related coordinate systems. The numerical scheme and especially its improvement by higher order shape functions is discussed in section 4, whereas in Section 3 we deal with the question of reliability and propose a method to represent this adequately in the resulting images. Furthermore in Section 6 we briefly give first results in 3D. Finally we draw conclusion and outline future research directions.

2 Lagrangian coordinates and transport equations

Velocity fields in numerical simulations are mostly given in the spatially fixed Eulerian coordinate system, whereas its physical meaning in terms of moving fluid particles is more closely related to the Lagrangian frame. This observation is the starting point of various visualization techniques. The method we propose here displays Lagrangian coordinates using a texture mapping, which map a certain pattern from a Lagrangian coordinates system to the Eulerian frame. To start with, let us assume $\Omega \subset \mathbb{R}^2$ to be a domain describing a fluid container with an inlet boundary $\Gamma^+ \subset \partial\Omega$ and an outlet boundary $\Gamma^- \subset \partial\Omega$. Furthermore we suppose the fluid velocity $v : \Omega \times [0, \hat{T}] \to \mathbb{R}^2$ to be given for a fixed time $\hat{T}$. In the application this velocity will be delivered by a numerical simulation, which runs simultaneously or has stored its results in files on disk. This numerical simulation is based on an additional computational grid. Therefore, to avoid some sampling procedure with its obvious drawbacks, the post processing method has to be based on the same grid (cf. Section 4).

Let us now interpret the coordinates X on the inlet boundary Γ^+, respectively the inflow time T as depending variables, which are transported with the fluid. Then they are described by the following transport equation for a density ρ

$$\partial_t \rho + v \cdot \nabla \rho = 0 \quad \text{in } \Omega,$$
$$\rho = \rho_\Gamma \quad \text{on } \Gamma^+, \tag{1}$$

thereby we obtain $\rho = X$ for $\rho_\Gamma = X$ on Γ^+, respectively $\rho = T$ for $\rho_\Gamma = T$ on Γ^+. At the outlet Γ^- no boundary condition has to be described if $v \cdot \nu \geq 0$ for all times, where ν is the outer normal of the domain Ω. This transport can be interpreted as a simultaneous and global particle tracing. On a particle path $x(t)$ the solution ρ of the above transport equation is constant, because $\dot{x}(t) = v(x(t), t)$ and

$$\frac{d}{dt}\rho(x(t), t) = \partial_t\rho(x(t), t) + \dot{x}(t) \cdot \nabla\rho(x(t), t)$$
$$= 0 .$$

Fig. 1. LIC convolution along the T component.

Therefore points of constant X value are located on the particle line starting at position X on Γ^+. Analogously a constant T value indicates points on a surface which is the image of a corresponding surface on the inlet under the flow $\phi(\cdot, T)$. In this sense X, T as functions on $\Omega \times [0, \hat{T}]$ can be regarded as Lagrangian coordinates describing the motion of particles which pass through Γ^+. Particles which have earlier entered the fluid container are not considered so far.

The transport equation becomes a well–posed problem by prescribing suitable initial conditions. If every particle paths starting at a position in Ω has left the domain, the solution ρ no longer depends on these initial conditions. For moderate values of $\hat{T}$ this might not be the case and for certain applications especially the initial phase of the physical simulation is of great importance. Therefore we suppose that $\tilde{X}$ and $\tilde{T}$ are extensions of $X|_{\Gamma^+}$ respectively 0 on Ω and choose them as initial conditions for the two transport problems. E. g. if $\Omega \subset \mathbb{R}^+ \times \mathbb{R}$ and $\Gamma_+ \subset 0 \times \mathbb{R}$ we choose $\tilde{X}(x_1, x_2) = (0, x_1)$, $\tilde{T}(x_1, x_2) = 0$.

In Section 4 we will discuss a numerical algorithm to compute an approximation of the transport solution and thereby of the Lagrangian coordinates.

Next we have to define an appropriate pattern in the texture space $\Gamma^+ \times [0, \hat{T}]$. There are several desirable features which should be realized by the textural representation of the Lagrangian coordinates. It should simultaneously code time and inlet coordinates. Furthermore to enable long time animation of moving fluids the pattern in the texture space should be periodic in T and the zooming into detailed areas has to be supported by a scaleability property. These requirements are fulfilled by the following construction:

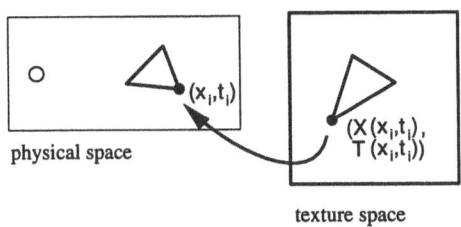

Fig. 2. A sketch of the applied mapping from texture space into physical domain Ω.

- Choose some white noise on a rastered domain $[0,1]^2$, those coordinates are denoted x, t corresponding to the Lagrangian coordinates X, T and duplicate this domain three times shifting it in the t-direction by $-1, 0$ and 1 (compare Fig. 1).
- Then use the LIC type convolution along the T component with a filter length smaller than 1 (compare Fig. 1).
- If a smoothing in x is intended repeat the same duplication and convolution in x direction with a second filter length. This mollification scale should be significantly smaller in order not to destroy the correct perception of the flow direction.
- Thereby we obtain a texture on the original domain $[0,1]^2$ with a previously fixed rasterization. By periodic shifting in both directions we finally obtain a 2 periodic texture.
- Depending on the projection from world to screen coordinates we scale the computed Lagrangian coordinates X and T by some factor λ. If λ_0 is an initial scale which especially depends on the size of the domain $\Omega \times [0, \hat{T}]$ and $s = (\det P)^{\frac{1}{3}}$ where P is the 3×3 projection matrix describing the linear part of the affine mapping from world to image space, then $\lambda := \lambda_0 \, s$ is an appropriate choice for this scaling factor.
- Finally we obtain as texture coordinates $\lambda X, \lambda T$ mapping points in Ω into the 2 periodic texture space $I\!\!R^2$ with the fundamental cell $[0,1]^2$ which covers $\{\lambda(X(x,t), T(x,t)) \,|\, x \in \Omega,\, t \in [0, \hat{T}]\}$ (compare Fig. 1).

Due to this construction the resulting texture on Ω at time $t \in [0, \hat{T}]$ continuously depends on t and the scaling from world space into image. Furthermore the resulting pattern is independent of this scaling. This avoids aliasing effects as long as the filter length in x direction is large enough.

Finally one degree of freedom is still left in the generation of an image. We can code by coloring a second scalar physical quantity, i. e. pressure. Alternatively color can be used to accentuate the motion on the streakline pattern. Therefore in addition a time periodic coloring of the greyscale texture is applied. In that case the T component of the Lagrangian coordinates is represented twice, by the periodic structure of the texture in T direction and by the coloring. If we disclaim the first we can abandon the T component of the texture to a reliability quantification of the numerical transport results. We will focus on this important aspect in the next paragraph.

3 Texture visualization and reliability

Although we use a higher order Finite Volume method to solve the transport equation for a given velocity v numerically, there are unavoidable error sources. In general, especially for CFD applications, v itself is computed by some numerical algorithms, which implies approximation errors compared to the true fluid velocity in the physical application and leads to errors in data v which we plug into the numerical transport scheme. Furthermore due to the still considerable numerical viscosity and the approximation restriction of the shape functions we obtain additional important errors contributions. Let us suppose that, by some error estimator [10, 18, 11] or a weaker error indicator we can measure local in space and time the resulting accumulated error. We will denote this measure $\eta(x, t)$ with $x \in \Omega$, $t \in \hat{T}$ and regard it as a function in the linear Finite Element space.

Our intention is now to use η information in the generation of the vector field images. In areas where η is small, the numerical solution of the transport equation and thereby the texturing of the domain Ω is reliable, whereas in regions with large η–values, the actual meaning of the texture is unclear and possible makes no sense.

Therefore we first create a texture π with a smooth transition from clearly visible pattern with a high signal bandwidth to an uniform grey valued texture. As already explained, if we code the Lagrangian coordinate T solely by color, the corresponding t texture component is no longer needed. Then we are able to parameterize the above transition over $t \in [0, 1]$. Let us suppose that the current one dimensional texture $\pi(\cdot, 0)$ is periodic with its fundamental cell $[0, 1]$, i. e. a white noise $\alpha(\cdot)$ on $[0, 1]$ periodicly expanded on $\mathbb{R}$ and finally convoluted by some block filter kernel $\chi_\epsilon(\cdot)$ with support ϵ. Here χ_ϵ denotes the characteristic function on $[-\epsilon, \epsilon]$. Then we have two methods at hand to define the required transition.

- We can expand the support of the filter kernel from ϵ at $t = 0$ to 1 at $t = 0$. In detail we define the texture at $t \in [0, 1]$ by $\pi(x, t) := \chi_{\epsilon(t)} * \alpha$ where $\epsilon(t)$ is a monotone function on $[0, 1]$ with $\epsilon(0) = \epsilon$ and $\epsilon(1) = 1$.
- Alternatively we can successively decrease the texture signal's amplitude. I. e. for given $\pi(\cdot, 0)$ and $\bar{\pi} = \int_0^1 \pi(x, 0) dx$ define

$$\pi(x, t) = (1 - \beta(t))\pi(x, 0) + \beta(t)\bar{\pi}$$

where β is a monotone increasing function on $[0, 1]$ with $\beta(0) = 0$ and $\beta(1) = 1$. In particular a spline β with vanishing derivatives at 0 and 1 has proved useful in the applications.

Finally these two methods can be combined by concatenation of the two operators (cf. Fig. 3 for the resulting texture).

With this new parameter family of one dimensional texture spaces at hand we now consider the implications of the error indicator on the choice of the actual texture coordinates. Let us suppose that η is a function with values in $[0,1]$, where small values indicate small error bounds and values closed to 1 large errors and therefore small reliability of the computational results. Then we take $(\lambda X, \eta)$ as texture coordinates which map the latter introduced texture onto the computational domain. Again this texture is scalable and continuously depends on time. The following examples for different applications all use this texture for the Lagrangian X coordinate and color for the corresponding T coordinate. Thereby a simple error indicator which measures local gradients has been applied.

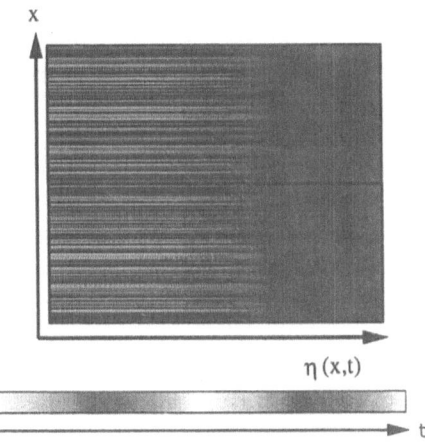

Fig. 3. Fundamental cell of the texture space with error dependent blurring and the periodic color ramp for the coding of time.

4   Higher order numerical transport scheme

Numerical schemes for hyperbolic conservation laws are accompanied by some numerical viscosity, which leads to a significant data mollification and a "smearing out" of the solution structure. This phenomena is well-known for shock propagation in CFD, but it already appears in case of linear transport problems. There is a trade off between the amount of this numerical viscosity and the occurring of oscillations. Especially in the current application to much numerical viscosity would destroy the evolution of interesting flow patterns represented in the numerical solution of our flow problem. Therefore, after some first testing we reject the usage of standard first order Finite Volume schemes and choose the higher order Discontinuous Galerkin method as an appropriate solver, with considerable smaller numerical viscosity.

The oscillations, which are well-known for any type of higher order finite volume scheme, are avoided by invoking a limiting process.

Let us suppose $\mathcal{M}$ to be some unstructured mesh covering the computational domain Ω and consisting of regular Elements E_i for i in some index set $I_{\mathcal{M}}$. On this grid we introduce the space $\mathcal{V}$ of piecewise polynomial function, which are not required to be continuous on element faces. Then we consider the transport equation (1), written in conservation form

$$\frac{\partial}{\partial t}\rho + \mathrm{div}\, f(\rho) = \rho \, \mathrm{div}\, v$$

where $f : \mathbb{R} \to \mathbb{R}^n$ and $f_i(\rho) := v_i \rho$, multiply it with some $\psi \in \mathcal{V}$ and integrate over $E \in \mathcal{M}$. Thereby we obtain

$$\frac{\partial}{\partial t}\int_E \rho\,\psi + \int_E \mathrm{div}\, f(\rho)\,\psi = \int_E \rho\,\mathrm{div}\, v\,\psi$$

Applying integration by parts we obtain

$$\frac{\partial}{\partial t}\int_E \rho\,\psi + \int_{\partial E} f(\rho)\cdot v\,\psi = \int_E \rho\,\mathrm{div}\, v\,\psi + f(\rho)\nabla\psi$$

If we now require $\rho \in V \times [0, T]$ and replace the flux term $f(\rho)\cdot v$, which describes the flow over the faces of E by some numerical flux $g(\rho^-, \rho^+)$ where ρ^- and ρ^+ denote the piecewise polynomial, but discontinuous function ρ in E, respectively in the adjacent cells $\tilde{E}$ at the faces of E, with

$$g(\rho, \rho) = f(\rho)\cdot v$$
$$g(\rho^-, \rho^+) = -g(\rho^+, \rho^-)$$

we obtain the semidiscrete Discontinuous Galerkin method. The Engquist-Osher flux [7] is used in the current texture transport algorithm. Finally we discretize this by some Runge Kutta scheme in time and to avoid oscillations combine the resulting algorithm with a limiter which cut off local extrema after each Runge Kutta iteration step. For a detailed discussion on the Discontinuous Galerkin Method we refer to Cockburn et. al. [3–5]. In our implementation we approximate $\rho = T, X$ on each volume E by a linear function. Let us emphasize that we obtain standard first order Finite Volume schemes if we take into account piecewise constant shape functions in space and a forward Euler scheme in time.

5 Examples in 2D

At first we considered an incompressible flow around a circular obstacle in a rectangular channel. At moderate Reynolds numbers we expect the Kármán vortex flow pattern. Here the numerical velocity v is calculated by a mixed Finite Element method with quadratic shape functions for v [1]. To resolve the

Fig. 4. A comparison between first and higher order method for the numerical transport of the X component using isolines on the physical domain Ω at a certain time.

approximation quality also in the numerical solver for the transport problem we refine the triangular Finite Element mesh, subdividing each triangle into 16 smaller triangles, and then start the second order Discontinuous Galerkin method to calculate the X, T coordinates. Several figures above already reflect the obtained results. Fig. 6 depicts several timesteps from the evolution of flow in time and Fig. 7 underlines the scalability of the texture for several different magnification factors. The zooming region is outlined in black in the original full image.

Finally we compute and display the texture transport for a compressible velocity field given by the numerical solution of the 2D Euler equations. Two obstacles are placed in a channel and we increase the prescribed inflow velocity successively in time. Fig. 8 compares the induced flow pattern at different times.

6 Vector fields in 3D

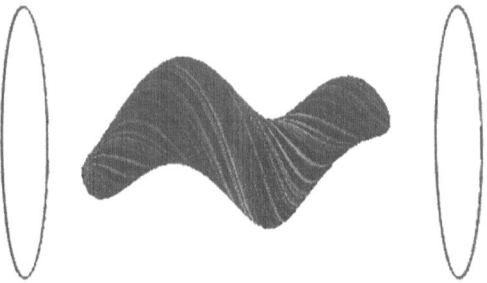

Fig. 5. 3D texture transport

The method of the Lagrangian coordinate transport can obviously be transfered to the three dimensional case. Thereby we especially compute the transport of two dimensional inlet coordinates $X \in \Gamma^+$. For a visualization of the results in terms of texture rendering, we pick up the implicit streamsurface idea presented by J. Wijk [23]. Consider implicitly defined curves $\gamma = \{X \in \Gamma^+ \mid g(X) = 0\}$ for some regular function g on Γ^+. Let us denote by $s : [0,1] \to \Gamma^+$ a parameterization of γ, which is supposed to be periodic if γ is closed. Then the image $X(\gamma, \cdot)$ of γ under the coordinate map X is a streamsurface. This surface can be extracted on the discrete grid by any discrete isosurface algorithm. With respect to the parameterization s it can be textured over the same texture space, already used in 2D applications. If we furthermore consider a family of implicit parameterized curves, we also obtain a continuous transition in the texture images concerning continuous modifications of this parameter in an interactive exploration. Instead of implicit curves on Γ^+ we can also ask for the images of implicit surfaces on $\Gamma^+ \times [0, \hat{T}]$ and texture them correspondingly. Fig. 5 shows a first picture of such a surface deformation by the Lagrangian coordinate mapping. Therefore an ellipsoid has been prescribed on $\Gamma^+ \times [0, \hat{T}]$. The underlying velocity is a test data set on a cylindrical domain.

7 Conclusions

A new method for the visualization of vector field data has been presented. It applies to stationary and time dependent data in 2D and combined with the implicit streamsurface method of van Wijk it has strong provisions for the three dimensional case as well. Based on the numerical solution of the transport equation for the Lagrangian coordinates (related to the inflow boundary) texture coordinates are calculated which map a pattern in the Lagrangian coordinate space onto the computational domain. The resulting pattern shows a strong alignment in the direction of particle paths and can be animated in time. The method is computationally expensive concerning the numerical solution of the transport problem, which may run in parallel to the actual numerical flow simulation or afterwards in a preparatory step for the post processing. Compared to this the actual post processing is fast and interactive especially on machines with hardware texturing. Future research will be on the distributed calculation of transport and the efficient extraction of well suited texture patterns in 3D. Here we will combine the proposed method with multilevel visualization techniques [14, 16, 15]. The authors thank E. Bänsch for providing the numerical data of the von Kármán vortex street.

This paper is a part of the PhD-thesis one author (J. Becker) is working on.

References

1. Bänsch, E.: Simulation of instationary, incompressible flows, Submitted to Acta Math. Univ. Comenianae.
2. Cabral, B.; Leedom, L.: Imaging Vector Field Using Line Integral Convolution, Computer Graphics Proceedings, Annual Conference Series 1993.
3. Chavent, G.; Cockburn, B.: The Local Projection P0-P1-Discontinuous-Galerkin Finite Element Method For Scalar Conservation Laws, Mathematical Modelling and Numerical Analysis Vol.23,N 4,1989,p.565-592.

4. Cockburn, B.; Shu, C.-W.: TVB Runge-Kutta Local Projection Discontinuous-Galerkin Finite Element Method For Conservation Laws II: General Framework, Mathematics of Computation, Vol.52,Nu.186,1989,p.411-435.

5. Cockburn, B.; Hou, S.; Shu, C.-W.: TVB Runge-Kutta Local Projection Discontinuous-Galerkin Finite Element Method For Conservation Laws IV: The Multidimensional Case, Mathematics of Computation, Vol.54,Nu.190,1990,p.545-581.

6. de Leeuw, W. C.; van Wijk, J. J.: Enhanced Spot Noise for Vector Field Visualization, presented at Visualization '95,Atlanta.

7. Engquist, B., Osher, S.: One sided difference approximations for nonlinear conservation laws. Math. of Comp. 36 (1981), 321-351.

8. Forssell, L. K.: Visualizing Flow Over Curvilinear Grid Surfaces Using Line Integral Convolution, IEEE Visualization '94,240–246, 1995.

9. Hultquist,J. P. M.: Interactive Numerical Flow Visualization Using Stream Surfaces. Computing Systems in Engineering, Vol.1, No.2-4, 1990, pp.349–353.

10. Kröner, D.: Numerical Schemes for Conservation Laws, Wiley Teubner.

11. Kröner, D.; Ohlberger, M.: A-posteriori error estimates for upwind finite volume schemes for nonlinear conservation laws in multi dimensions. Preprint, Mathematische Fakultät, Albert-Ludwigs-Universität Freiburg. 1998.

12. Max, N.; Crawfis, R.; Grant, C.: Visualizing 3D Velocity Fields Near Contour Surfaces, IEEE Visualization '94,248–254, 1994.

13. Max, N.; Becker, B.: Flow Visualization using Moving Textures, Proceedings of the ICASE/LaRC Symposium on Time Varying Data, NASA Conference Publication 3321, D. C. Banks, T. W. Crocket, and K. Stacy, editors,(1996) pp. 77 - 87.

14. Neubauer, R.; Ohlberger, M.; Rumpf, M.; Schwörer, R.: Efficient Visualization of Large–Scale Data on Hierarchical Meshes. Lefer, W. and Grave, M., Visualization in Scientific Computing, 1997, Springer.

15. Ohlberger, M.; Rumpf, M.: Adaptive Projection Methods in Multiresolutional Scientific Visualization. Report 20, Sonderforschungsbereich 256, Bonn, 1998.

16. Ohlberger, M.; Rumpf, M.: Hierarchical and Adaptive Visualization on Nested Grids. Computing. Vol.59 (4), pp.269-285, 1997.

17. SFB 256, University of Bonn: GRAPE manual, http://www.iam.uni-bonn.de/main.html, Bonn 1995.

18. Sonar, T.; Süli, E.: A dual graph-norm refinement indicator for finite volume approximations of the Euler equations. Oxford University,Report 94/9 , 1994.

19. Stalling, D.; Hege C.: Fast and Resolution Independent Line Integral Convolution, Proceedings SIGGRAPH '95, 1995.

20. Stalling, D.; Zöckler; M.; Hege, H.-C.: Fast Display of Illuminated Field Lines. IEEE Transactions on Visualization and Computer Graphics, No. 2 Vol.3 1997.

21. Turk, G.: Re-tiling polygonal surfaces. Computer Graphics (SIGGRAPH '92 Proceedings) Vol.26 1992,55–64.

22. van Wijk, J. J.: Spot Noise – Texture Synthesis for Data Visualization, Computer Graphics, Volume 25, Number 4, 1991.

23. van Wijk, J. J.: Implicit Stream Surfaces, IEEE Visualization '93,245–252, 1993.

24. van Wijk, J. J.: Flow Visualization with Surface Particles. IEEE Computer Graphics and Applications Vol. 13, No.4, 1993 , pp.18–24.

Editor's Note: see Appendix, p. 148 for colored figures of this paper

Part IV
Visualization Quality

Stereoscopic Volume Rendering

Roger J. Hubbold[1], David J. Hancock[1], and Christopher J. Moore[2]

[1] Advanced Interfaces Group,
Department of Computer Science,
University of Manchester,
Manchester M13 9PL
[2] Department of Medical Physics,
Christie Hospital,
Manchester M20 9BX

Abstract. In this paper we describe the extension of a parallel, distributed, direct volume renderer for use with a novel auto-stereoscopic display. We begin by briefly describing the target application of our research, radiation therapy planning, why we believe that stereoscopic viewing may be helpful for this, and the design of our DVR system. We then report on some of the problems we have encountered, and the results we have obtained in experiments. These demonstrate that stereoscopic viewing is beneficial for perceiving depth in transparent DVR images. We illustrate the application of the system to the visualisation of prostate cancer treatment plans. Finally, we describe the use of head-tracking to implement 3D stereo look-around.

Keywords: auto-stereoscopic display, volume rendering, visualisation, medical imaging, radiotherapy.

1 Introduction

For several years, direct volume rendering (DVR) by ray casting has been investigated as a tool for visualisation in medicine [11]. Here, we report on experiments relating to the application of DVR to radiation therapy (radiotherapy) planning. Radiotherapy planning is a safety-critical task – any mistakes could have serious consequences for the patient. It is also difficult, because it is a complex 3D problem involving multi-valued volume data. Our goal is to find ways to visualise treatment plans using 3D displays. These 3D views must unambiguously convey all of the information needed to ensure that a plan is viable and safe.

One approach to disambiguating depth in DVR images is to use a stereoscopic display. We are fortunate in having a collaboration with the Imaging Technology Group at Sharp Laboratories of Europe. They have developed a number of stereoscopic display technologies, including auto-stereo devices in which no special glasses, filters or shutters need be worn. The device which we have been testing presents a full-resolution image to each eye simultaneously [4] – an important consideration when high-quality images are needed for medical visualisation. Using this device, we have conducted experiments to evaluate whether stereoscopic visualisations offer advantages compared with monoscopic ones.

Direct volume rendering has been proposed previously for radiotherapy planning [12, 5]. However, we believe that our work is the first attempt to apply the method to conformal radiotherapy, and to investigate the use of auto-stereoscopic displays for this purpose.

The remainder of this paper is organised as follows. We begin with a brief overview of the target problem, radiation therapy planning. Next, we outline our parallel, distributed DVR system, and explain the techniques we use to obtain interactive rendering speeds. We then consider the problems of extending these ideas to work with an auto-stereoscopic display. We summarise results of experiments to explore whether stereoscopic display is helpful in interpreting DVR images. We show that although DVR is helpful, some of the techniques which can be used for monoscopic images do not work satisfactorily in stereo, because of aliasing problems. We illustrate the application of our method with some visualisations of a treatment plan. Finally, we describe the use of head tracking to control viewing on the stereo display.

2 DVR visualisations of radiation therapy plans

The purpose of radiation therapy planning is to devise a pattern of radiation beams to irradiate a cancer within a patient's body. In our study the cancers are in the prostate, which is surrounded by sensitive organs such as the bladder and rectum. The directions of the different beams can be adjusted relative to the patient's body in order to target the tumour. The cross-sectional profiles of the beams can also be varied using a series of programmable tungsten leaves – a process known as *conformal radiotherapy* [9, 2]. The goal is to plan multiple intersecting beams, which enclose the tumour. In this central region, where the beams intersect, a high radiation dose is delivered, but this must be achieved without damaging other organs. Current clinical practice is to use a series of 2D slices for visualising plans, but this has the serious drawback that a 3D model of the plan is only formed in the consultant's head, rather than visualised directly.

In order to determine whether a given plan is satisfactory, several different data fields must be combined: CT or MRI scans of the patient, the tumour volume, the geometry and magnitude of the radiation dose field (computed using a procedure which takes account of propagation and scattering), and particular organs which may be sensitive to radiation damage, such as the rectum and bladder. In our experiments the tumour and organs are segmented during a preprocessing step, so our task is to produce a coherent view in which different features are clearly visible.

The dose field has a complex 3D geometry, which makes it difficult to visualise. The field strength varies continuously through the volume, adding to the difficulty. The dose field is only really useful when viewed in combination with other features, such as the tumour volume. For example, we must be able to visualise the radiation dose inside a tumour to ensure that an adequately high level is delivered to the target region. Too low a dose results in a *cold spot*. Conversely, if the dose is too high inside critical regions, such as the rectum, then irreparable damage can result; such occurrences are termed *hot spots*. Because these hot and cold spots occur *inside* organs, some means is needed for seeing through the data. DVR offers a number of potential advantages for this, as volumes can be made partially transparent. However, our experience with monoscopic, transparent DVR images shows that depth relationships – important in our application – can be very difficult to discern. This provides the motivation for exploring whether stereoscopic renderings are helpful in interpreting the data correctly.

3   The KVR direct volume renderer

We use a purpose-designed, parallel, distributed volume renderer called KVR. We have previously reported a number of design issues and experimental results relating to this. Here, we restrict our discussion to features which impact – in some cases indirectly – on the effectiveness of stereoscopic display.

Direct volume rendering is notoriously time consuming, but when implemented carefully is capable of generating high-quality images with minimal artifacts. To address the computation speed we use a parallel implementation running on a remote supercomputer [6]. In order to access this from low-cost workstations and PCs we developed a distributed rendering protocol [7]. Its design is integrated with the parallelisation strategy of the renderer, so that images are compressed, transmitted, and displayed using a pipeline design to hide network latencies. Furthermore, the implementation includes a progressive refinement strategy so that initial low-resolution images are delivered and displayed very rapidly [6]. An initial image is displayed in a fraction of a second, and a 640×480 full-resolution image (one ray per pixel) takes five seconds to generate on 24 processors on an SGI Origin-2000[1].

This integration of parallel rendering, progressive refinement and pipelined design works very effectively for monoscopic images. Closed-loop times for interaction of less than one second make it feasible to perform direct manipulation of the visualisation parameters (viewpoint, lighting, classification, cutting planes and sub-volumes). Extending the approach to work with stereoscopic displays appears, at first sight, to be straightforward. However, in practice this has not proved to be the case.

4   Extending DVR for stereoscopic display

The most obvious change required for stereoscopic display is the generation of two views, one for each eye. Our Sharp auto-stereo display (ASD) is driven

---

[1] This time was actually measured for a *pair* of stereo images; the time for a single monoscopic image is actually half of this.

from an SGI Crimson/VGXT via a videosplitter option. Each view is composed in a separate window on the SGI screen and these are fed as standard NTSC signals into the display hardware. Within the display, each view is presented on a dedicated LCD panel, and the two images are combined using a patented system which ensures that each eye sees only the correct image. The display's design ensures that a full 640×480 image is presented continuously to each eye, without flicker and with minimal cross-talk. With some types of auto-stereo display, multiple views are obtained by trading horizontal resolution against the number of views. Not only does this lead to lower effective horizontal resolution, but it introduces a correspondingly lower depth resolution. With the Sharp ASD, because each image is shown at full resolution, fine detail is visible and excellent depth resolution is achieved.

Such high quality is important for applications such as medical imaging, but equally means that the renderer must generate images of comparable fidelity; any unwanted artifacts will be uncompromisingly revealed by the display. Although many systems have been used for stereoscopic display, image quality has often been compromised by lower resolution, or by cross-talk between the left- and right-eye views. The human visual/perceptual system is remarkably adept at attempting to compensate for such deficiencies, but prolonged use of inadequate displays leads to visual fatigue. In the literature on stereoscopic displays for visualisation these issues have received scant attention.

In our system the left and right views are generated using the standard technique of parallel, perspective projections [8]. The offsets used for the eye positions and 'look points' are calculated to match the view plane distance of the ASD. This yields a correct perception of depth for the presented images. A novel feature of our system is that the user interface and renderer are designed to support an arbitrary number of stereo visualisations of a given data set. Each stereo pair can be classified and coloured to show different features within the volume data. These multiple visualisations can then be composed into a single 3D normalised projection coordinate space which is projected stereoscopically to yield a correct 3D view. Figure 5 shows an example.

It is important to realise that this is *not* the same as generating several stereoscopic pairs of images and then compositing these in 2D – this latter approach leads to anomalous views, resulting in distortions during viewing (see Section 8). Because we are trying to make important decisions about relative positions of features, accurate display is essential. We do not know of any other stereoscopic DVR systems which can present multiple views in this way. Being able to view several complementary, stereoscopic visualisations simultaneously in this way has proved to be a very effective technique.

5   Experiments to assess stereoscopic DVR

In an earlier paper [10] we described a set of experiments to quantify the value of stereoscopic display for a synthetically generated set of test volumes. In this section we briefly summarise the most important results, and then in the next section we consider how these have affected our use of stereoscopic display for the radiotherapy application.

5.1 Depth tests

A first experiment was designed to test whether users could differentiate very small depth differences in a series of images. The test images were generated from a *voxelised* dataset containing three small spheres enclosed in a larger transparent sphere. Three different types of image were generated by varying the classification of the data volume. They were selected to be representative of the principal effects obtainable with DVR – combinations of surface shading and transparency. Figures 4 (a) – (c) show examples of different classifications. In each case, the small spheres were made totally opaque, as we wished to assess the effect of visualising these inside another volume.

In the first classification ("Inner-only", Figure 4 (a)), the large outer sphere was made totally transparent, so that only the small spheres were visible. This provided a base case for reference, although it made judging depth very difficult. This was because there were no depth cues other than those resulting from very small variations in size due to perspective. In the next version ("Transparent", Figure 4 (b)), the outer sphere was made partially transparent, but surface effects were suppressed. The effect of this is to produce an attenuation of brightness of the inner spheres. In the third classification ("Shell", Figure 4 (c)), the surface of the outer sphere is enhanced using gradient magnitude. This produces an image in which the outer surface has a shell-like appearance.

For each classification, images were generated in which the central, small sphere was stationary, but the two outer, small spheres were moved either forwards or backwards by a small amount. If the left sphere was moved forwards, the right one was moved backwards by a corresponding amount, and vice-versa. The amount of movement was deliberately made very small, so that differences between images were very subtle. For each combination of classification and sphere positions, both monoscopic and stereoscopic views were generated.

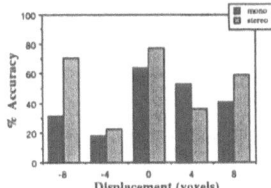 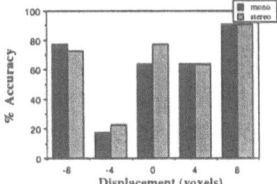 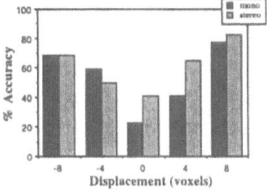

Fig. 1. Inner-only   **Fig. 2.** Transparent   **Fig. 3.** Shell

The resulting thirty images were shown in a random order to 22 subjects, and they were asked to choose one of three possible answers: left sphere is closer, right sphere is closer, or all three spheres have equal depth. Subjects were not told whether a given image was monoscopic or stereoscopic.

The results, shown in Figures 1 to 3, demonstrate a clear advantage in favour of stereoscopic display. The best results were obtained in the third category, in which the outer sphere's surface was accentuated. This is perhaps not surprising, as stereo helps to accentuate depth perception when clearly delineated features are present. Most surprising was the fact that stereo was most helpful in disambiguating situations where other effects were very small. For example, in the case where all three spheres were at the same depth, the results were better with stereo.

5.2   Aliasing problems

Two types of aliasing affect stereoscopic images. These are non-correspondence aliasing and depth aliasing [13]. Non-correspondence aliasing results when features which should be visible to both eyes appear in only one image of the stereo pair. In DVR this is caused by a ray completely missing the feature in one of the views. Depth aliasing results from the discrete number of apparent depths which can be displayed for a given horizontal pixel pitch.

In common with other systems, in our earlier work we relied on progressive refinement to deliver interactive image generation times for monoscopic images. Unfortunately, under-sampling of the data to obtain initial low-resolution images does not work in stereo. This is because ray casting is a point sampling technique. If only a small number of rays are cast, then discrepancies arise between the left and right images. Features may, and often do, appear in one image and not the other, and this can cause considerable discomfort for the viewer. Aliasing is a serious problem for good quality stereoscopic display, but is often ignored. During brief periods of use the problem may go unnoticed and is often not even considered. But aliasing makes correct depth perception difficult (conflicting information is presented to the two eyes), and prolonged use results in visual fatigue.

We have found little published data dealing with this topic for volume rendering, although Adelson and Hansen [1] referred briefly to the problem. Their solution was to re-project a feature intersected by a ray in one view into the other view. They showed how this could yield accelerated rendering by reducing the work needed to obtain two views. However, it is clear that the method described in their paper will result in depth aliasing.

Sampling theory tells us that to avoid aliasing we must sample data at a rate above the Nyquist frequency. In DVR, this frequency must be chosen so that the spacing between rays, and the distance between re-sampling points along each ray, is less than the distance between data points (voxels). One possible way to address this problem is to pre-filter the volume to produce low resolution datasets, and to use these to generate initial images during progressive refinement.

To test this assertion, we computed filtered, lower-resolution versions of the full-resolution volume. We then generated images corresponding to three different levels of progressive refinement. Three images were generated with the full resolution volumes, and a further three using the lower resolution volumes which best matched the image resolutions. Subjects were shown the six different images and asked to assign two scores to each image, one for their perception of depth, and the other for viewing 'comfort'. Scores for both measures were assigned on a scale from 0 (very bad) to 4 (very good).

image resolution	volume resolution full	matching	% benefit
low	1.43 ± 0.87	2.33 ± 0.96	63%
medium	2.31 ± 0.90	3.26 ± 0.78	41%
high	2.62 ± 1.24	3.52 ± 0.62	34%

Table 1. Scores for perceived depth

image resolution	volume resolution full	matching
low	2.04 ± 1.17	2.09 ± 1.06
medium	2.43 ± 0.86	3.18 ± 0.65
high	3.61 ± 0.60	3.57 ± 0.78

Table 2. Scores for viewing comfort

Tables 1 and 2 show the scores (along with the standard deviations) allocated for comfort and depth respectively to each of the six images. The results confirm that depth perception improves with the sub-sampled volumes which correctly match the image resolution. The improvements are proportionately greater at lower resolutions. However, the scores assigned for 'comfort' showed only a minor improvement at the low and medium resolutions. From subjects' comments, it was clear that there was some confusion between comfort and image resolution. They strongly preferred looking at higher resolution images, but there was no clear favouring of the correctly sub-sampled volumes.

6   Implications for stereo DVR

These results have two important consequences for interactive DVR. First, stereoscopic viewing yields a demonstrable improvement in depth perception for volume rendered images containing nested transparent objects. Because DVR images can take some time to generate, this benefit may be quite important in our medical imaging application.

Second, although the effects of aliasing can be mitigated by using multiresolution volumes, depth aliasing cannot be removed because it results from pixel replication. Users are not happy looking at low resolution images in stereo because of correspondence problems. Given that DVR uses ray casting, which is a point sampling technique, there does not appear to be any easy solution to this problem. Using filters, or tri-linear interpolation, helps, but does not remove the problem. The practical consequence of this is that during progressive refinement, monoscopic images should be used. This is easily arranged by displaying a single image for both eyes, even though both images are computed. Once full resolution is reached, the correct second view can be switched in, giving the full benefits of extra depth information. Although not ideal, this is a practical compromise, which gives the benefits of rapid closed-loop interaction, and stereoscopic viewing of complete images.

7 Visualisation of radiotherapy plans

Figures 5 and 6 show examples of the volume renderer applied to a particular set of patient data. These images need to be viewed in colour, and may be accessed at http://www.cs.man.ac.uk/aig/ASD/.

In Figure 5 we show four different visualisations composited into a single view. For clarity, we show this monoscopically, although our system generates a correct stereo rendering of this picture. The upper-left picture shows the dose field, coloured red, resulting from four treatment beams. Organs are coloured green, yielding a yellow colour where the two intersect. A bright yellow indicates a high dose. The bottom left image shows the corresponding hot (red) and cold (blue) spots. In the bottom-right image, we have added the organs (green). Finally, in the top-right image, we have added the pelvic girdle, obtained from the original CT scans. The option to produce a number of visualisations showing different classifications of the data, correctly composited for stereo viewing, greatly aids our ability to relate the different features to each other.

Figure 6 contains an example of a stereo rendering. The two side-by-side images are arranged for cross-eyed viewing (with the right-eye view on the left). If you are able to fuse these images it should be evident that the stereo view gives a much clearer interpretation of the depth than looking at either image individually. We should emphasise that on the Sharp auto-stereo display, these images can be viewed very comfortably. In this picture, we see several nested features. The tumour is coloured green. This is surrounded by the 'planning tumour volume' (PTV), coloured light grey. The PTV is the region which the clinicians have decided should receive a high dose, to allow for the possibility of movement of the organs during a course of treatment. This is therefore the target region. In addition, we see the bladder (the somewhat rectangularly shaped blob towards the bottom of the image) and a section of the rectum (the tube-shaped feature at the top). The white lines represent the centres of the radiation beams. As previously, hot and cold spots are shown.

8 Use of head tracking to control viewpoint

An alternative, complementary technique for depth perception is the use of rotation. This can take a number of forms: direct user control of motion, automatic 'rocking' of the image between two angular positions, and – where suitable hardware is available – head tracking.

The Sharp ASD has a head tracking facility. In normal use, this is coupled to a unique image steering mechanism which ensures that the left- and right-eye images remain correctly aligned with the user's eyes. An important feature of this is that a full resolution image is continuously available to each eye, unlike other ASD systems in which horizontal resolution is reduced to obtain stereoscopic views and/or look-around capability. The Sharp system also provides the option to feed the user's head position into the host computer. This means that it can be employed to change the viewing parameters. With a little calibration, it is possible to provide a 'look-around' capability, in which the 3D objects appear stationary in front of the viewer as the head is moved from side to side. Without this, a lateral movement of the head produces a strange apparent distortion of the displayed view – this is the effect referred to previously in Section 4.

To assess this technique, we have generated sequences of full resolution, stereoscopic images (such as those shown in Figure 5) to cover a range of viewing positions in small angular increments. In look-around mode, the view displayed is chosen to correspond to the user's head position, and this is updated dynamically. To provide a comparison, a rocking mode has also been implemented, in which the view oscillates at a user-controllable rate. With both methods of control, the display can be switched between stereoscopic and monoscopic viewing.

Preliminary results from this once again provide convincing evidence of the value of stereoscopic display. Although the rocking motion gives a clear sense of depth, continuous movement makes it impossible to study the data carefully. For this, rocking must be temporarily suspended. In look-around mode, it is possible to freeze the view simply by keeping the head reasonably still. Changing view is intuitively easy and feels quite natural. Distortions which would otherwise result from head motion are eliminated. If stereoscopic viewing is enabled, then depth perception persists when motion is stopped. However, if monoscopic viewing is used then the image immediately becomes 'flat' and depth interpretation is difficult.

The combination of head tracking and stereoscopic viewing is very attractive. However, it is not cost-free. As noted previously, monoscopic presentation should be used during progressive refinement. Because a full resolution image takes several seconds to generate, the look-around mode is only feasible with pre-computed images. Also, there is an associated cost for the necessary hardware in the display itself. We plan to conduct a more complete set of tests with a representatively large sample of users to better quantify the benefits of this technique.

9   Concluding remarks

We have demonstrated that there are important benefits in using stereoscopic presentation for DVR. Our experiments with depth perception show a clear advantage for stereo, even in cases where depth relationships are very subtle, and where transparency makes judgement difficult. However, several problems arose, of which the most important are non-correspondence and depth aliasing. These arise particularly with progressive image refinement, during which we revert (temporarily) to monoscopic viewing.

The problems of aliasing should not be underestimated with stereoscopic displays. They are frequently ignored because the human visual/perceptual system attempts to make sense of anomalous images, but they are present nonetheless, and for good quality images anti-aliasing must be used. Without this precaution, visual fatigue – evident in our multi-resolution aliasing experiments – will result, making stereo displays difficult to view for extended periods.

There are, of course, other approaches to volume rendering. Candidates of particular relevance for the kinds of visualisations with which we are concerned here are splatting [14], and 3D texture mapping (α-blending) [3]. The latter, in particular, has the potential to generate images quite rapidly using modern display hardware. It would be instructive to investigate the extent to which these methods suffer from aliasing problems when used with a high-quality stereoscopic display.

Acknowledgements

This research is funded by the UK's Engineering and Physical Sciences Research Council under grant GR/L02685. The auto-stereoscopic display was developed, and lent to our group, by the Imaging Technology Group of Sharp Laboratories of Europe, Oxford, UK. We are very grateful to Dr David Ezra and his staff, and to Sharp Laboratories, for their support.

References

1. S.J. Adelson and C.D. Hansen. Fast steroscopic images with ray-traced volume rendering. In Arie Kaufman and Wolfganag Krueger, editors, *1994 Symposium on Volume Visualization*, pages 3–9, Washington D.C., October 1994. ACM Press.
2. A.R. Hounsell et al. Dose calculations in multileaf collimator fields. *British Journal of Radiology*, 65:321–326, 1992.
3. Brian Cabral, Nancy Cam, and Jim Foran. Accelerated volume rendering and tomographic reconstruction using texture mapping hardware. In Arie Kaufman and Wolfganag Krueger, editors, *1994 Symposium on Volume Visualization*, pages 91–98, Washington D.C., October 1994. ACM Press.
4. David Ezra, Graham J. Woodgate, Basil A. Omar, Nicolas S. Holliman, Jonathan Harrold, and Larry S. Shapiro. New autostereoscopic display system. In *Proc. SPIE 1995 Conference*, volume 2409, pages 31–40, 1995.
5. D. Gordon, M.A. Peterson, and R.A. Reynolds. Fast polygon scan conversion with medical applications. *IEEE Computer Graphics and Applications*, 14(6):20–27, November 1994.
6. D.J. Hancock and R.J. Hubbold. Distributed parallel volume rendering on shared-memory systems. In Bob Hertzberger and Peter Sloot, editors, *High-performance computing and networking*, number 1225 in Lecture Notes in Computer Science, pages 157–164. Springer-Verlag, April 1997. (Proc. HPCN Europe 1997. ISBN 3-540-62898-3.).
7. D.J. Hancock and R.J. Hubbold. Efficient image synthesis on distributed architectures. In Rae Earnshaw and John Vince, editors, *The Internet in 3D, Information, Images and Interaction*, pages 347–364. Academic Press, 1997. ISBN 0-12-227736-8.
8. Larry F. Hodges and David F. McAllister. Computing stereoscopic views. In David F. McAllister, editor, *Stereo Computer Graphics and Other True 3D Technologies*, chapter 5, pages 71–89. Princeton University Press, 1993.
9. A.R. Hounsell and J.M. Wilkinson. Dose calculations in multileaf collimator fields. In *Proc. XIth International Conference on Use of Computers in Radiation Therapy*, pages 234–235. Manchester, UK, 1994.

10. R.J. Hubbold, D.J. Hancock, and C.J. Moore. Autostereocopic display for radiotherapy planning. In Scott S. Fisher, John O. Merritt, and Mark T. Bolas, editors, *Stereoscopic Displays and Virtual Reality Systems*, volume 3012, pages 16–27. SPIE – The International Society for Optical Engineering, February 1997.

11. Arie E. Kaufman. Introduction to volume visualization. In Arie E. Kaufman, editor, *Volume Visualization*, chapter 1, pages 1–18. IEEE Computer Society Press, 1991.

12. M. Levoy. A hybrid ray tracer for rendering polygon and volume data. *IEEE Computer Graphics and Applications*, 10(2):33–40, March 1990.

13. William F. Reinhart. Gray-scale requirements for anti-aliasing of stereoscopic graphic imagery. In *Proc. SPIE 1992*, volume 1669, pages 90–100, 1992.

14. L. Westover. Footprint evaluation for volume rendering. *ACM Computer Graphics*, 24(4):367–376, August 1990.

Mirror, Mirror on the Wall, Who Has the Best Visualization of All? – A Reference Model for Visualization Quality –

Helmut Haase

Fraunhofer Institute for Computer Graphics (IGD),
Rundeturmstrasse 6, 64283 Darmstadt, Germany
http://www.igd.fhg.de/~haase/, Email: haase@igd.fhg.de

Abstract. What is a 'good' visualization, one which leads to desired insights? How can we evaluate the quality of a scientific visualization or compare two visualizations (or visualization systems) to each other?
In the following, the importance of considering the 'visualization context' is stressed. It consists of the prior knowledge of the user; the aims of the user; the application domain; amount, structure, and distribution of the data; and the available hardware and software. Then, six subqualities are identified: data resolution quality, semantic quality, mapping quality, image quality, presentation and interaction quality, and multi-user quality. The Q_{VIS} reference model defines a weight value C (i.e., importance) and a quality value Q for each subquality. The Q_{VIS} graph is introduced as a compact, easy to perceive representation of the so-defined visualization quality. An example illustrates all concepts.
The reference model and the graph can help to evaluate visualizations and thus to further improve the quality of scientific visualizations.

1 Introduction

In the fairy-tale 'Snow White', the wicked stepmother asks the mirror: 'Mirror, mirror on the wall, who is the loveliest lady in the land?' — By what criteria does the mirror evaluate the 'loveliness' of women? Surely, 'loveliness' cannot be measured objectively. Rather, loveliness will be defined differently by different people. Similarly, in order to evaluate the quality of a scientific visualization, it is important to know who will use this visualization and for what purpose.

Good scientific visualizations are needed by scientists and engineers in many fields, but they can also be useful to managers and to the general public. Here, 'good' generally means 'meaningful', but also 'beautiful' (by artistic standards) or 'simple' visualizations can be desirable. In this work, 'visualization quality' will mainly be examined from a technological point of view and psychological, pedagogic, or artistic factors mostly have to be excluded.

Any dataset can be visualized in an infinite number of possible ways. If the visualization is to achieve its goal (e.g., lead to new insights quickly), the choice of a visualization system, of visualization techniques, as well as of visualization parameters is crucial. But what is a 'good' visualization? How well does the result of a visualization (i.e., an image or an animation) or a visualization system reach its goal? How can we compare strengths and weaknesses of such visualizations or systems? – In the following, a reference model for 'visualization quality' will be presented as a foundation for discussing such questions.

Another question is: Can an 'objective' measure be achieved at all? Surely, 'visualization quality' always depends on a 'visualization context' which includes (but is not limited to) the prior knowledge and the aims of the user.

The importance of evaluating visualization software has been stressed by Globus and Uselton in [12]. There, a number of possible evaluation methods has been proposed, ranging from the analysis of mathematical properties of algorithms to performance measurements of users. On the other hand, Robertson and Silver [29] recommend case studies. They point out that in a specific application case, it is more easy to decide if the goals of a visualization have been met and how an increased effectiveness, reliability and consistency of visualizations can be achieved over a wide range of application domains. [3] gives one example of such a case study. Here, several ways of visualizing a storm are presented and evaluated. The discussion of the quality of these visualizations also takes into account graphics design and perceptive issues.

An extensive selection of visualization examples may be found in the book 'Visual Cues' [23]. Each of them is described in picture and text on one page. These examples are ordered according to classes of visualization techniques. The number and variety of examples allow a good comparison and an evaluation of visualizations. Several appendices give fast access to these examples according to visualization techniques, visualization goals, number of visualized variables, application domains or the used hard- and software. A theory chapter of the same book gives general hints for good visualizations, including visualization goals, output media, design principles, and usage of color.

In the field of automatic generation of visualizations, much important work relevant to the question of visualization quality has been done. Starting from the work by Mackinlay [24], who introduced the two criteria of expressiveness and effectiveness, there has been a considerable number of interesting works, including [27], [2], [20], [25], [28], [4], [26], [30], and [22].

2 The Q_{VIS} Reference Model for Visualization Quality

2.1 Short Definiton of Visualization Quality

Before starting to explore the 'measurement' of visualization quality, here is a definition of how the term is understood in this paper.

The quality of a visualization is defined as: the possibility and ease for a specific user to gain the insight desired by him into information that is conveyed in his data by looking at or interacting with the visualization.

The quality of a visualization *system* is defined as: the possibility and ease for an average user from a clearly defined group of users with a clearly defined goal in a clearly defined application context to gain the desired insight into information that is conveyed in an average data set of a clearly defined set of possible data sets by using the visualization system in a clearly defined hardware and software environment.

What influences the so-defined 'visualization quality'?

2.2 Visualization Context

The work reported, e.g., in [8], [15], [14], [17], and [13] has confirmed that we cannot speak of the 'quality of a visualization' without considering the environment in which a visualization occurs, i.e., its application context. We call this the 'visualization context'. It includes:

1. the prior knowledge of the user,
2. the aims of the user,
3. the application domain,
4. amount, structure, and distribution of the data,
5. the available hardware and software.

In [9], a similar scheme is proposed. Due to space limitations, please refer to [16] for a more detailed explanation of the visualization context[1].

2.3 Visualization Subqualities

Six 'subqualities' together describe visualization quality in the Q_{VIS} reference model:

1. data resolution quality (dr),
2. semantic quality (se),
3. mapping quality (ma),
4. image quality (im),
5. presentation and interaction quality (pi),
6. and multi-user quality (us).

The *data resolution* depends on the number of data values in relation to the given range and to the underlying function they sample. If the quality of a visualization *system* is under investigation (as opposed to the visualization quality for a specific dataset), then the subquality for data resolution should not be considered.

The *semantic quality* stands for the semantics of the data to visualize. Four cases can be distinguished: no semantics (i.e., geometry, color, etc. only, not derived from original values); static semantics (geometry, etc. derived from static data); offline-dynamic semantics (geometry, etc. derived from initially known,

---

[1] In [16], the term 'visualization background' was introduced by the author, but 'visualization context' better describes the concept and therefore will be used from now on.

dynamic data); and online-dynamic semantics (geometry, etc. derived from on-line simulation or online measurement, i.e., the data is being generated concurrently with the process of visualization, e.g., as response to interactive steering of the user). Thus, semantic quality comprises the degree of direct interaction of the user with the data source.

The *mapping quality* is the next important subquality. it includes the flexibility of mapping original values to visualization objects, the numerical quality of this process (interpolation, integration in vector fields, etc.), and the consideration of human perception, e.g., in the case of color selection.

The *image quality* mainly includes five subitems: image resolution (number of pixels), color space resolution, dynamic range, pixel sharpness, and rendering quality. Most of these subitems need not be explained here since they are discussed in many publications, e.g., in [7]. The subitem rendering quality (in the 3D case) distinguishes different rendering techniques like wireframe, flat shading, Phong shading, raytracing with reflections, etc. Thus, image quality in this paper only comprises technical, static image quality. Content or aesthetics are either partly covered by other subqualities (e.g., data resolution or mapping quality) or completely excluded from this reference model.

The *presentation and interaction quality* includes: temporal resolution (frames per second), latencies, field of view, stereoscopic quality, degree of immersion due to head tracking, and intuitivity of input devices. Thus, this subquality comprises all kinds of presentation and interaction starting from batch processing (latency, e.g., one day), interactive graphics (1 to 10 frames per second, latency less than 1 second) to immersive visualization (typically more than 10 frames per second, latency less than 0.2 seconds). Some of the subitems of this subquality are mainly interesting for immersive visualization[19] (field of view, immersion due to head tracking) while others are important for many more visualization applications (temporal resolution, latencies, stereoscopy).

Finally, the *multi-user quality* takes account of the number of users of a visualization (system). It distinguishes users that are interacting online with the visualization, users that are consuming online (but without interaction), and users that are consuming offline, i.e., they see the results of a visualization process after the visualization has been completed. The number of online interacting users again can be grouped in four important classes: no interacting user during the generation of a visualization (i.e., batch processing), one interacting user (most common case), two interacting users (simple CSCW[2] solution for connection and consistency), and more than two interacting users (complex CSCW connection and consistency structure). Another aspect in respect to multi-user quality is the location of users: Do all of them have to be in the same room or can they be distributed over large distances?

---

[2] CSCW = Computer Supported Cooperative Work

2.4 The Q_{VIS} Reference Model

The reference model for visualization quality defines a way to get numerical quality values for a visualization by quantifying a number of subqualities as well as their importance (as 'weight values') under consideration of the visualization context.

Each subquality is quantified by determining a *subquality value* in the range from 0.0 to 1.0. These subquality values are also denoted by Qxx (Qdr, etc.). A subquality value expresses how well the visualization under investigation satisfies the demands of the visualization context in the corresponding subquality.

One 'weight value' Cxx (Cdr, etc.) is assigned to each of the subqualities in order to express the importance of the subquality for the overall task (again depending on the visualization context). This is necessary since not all subqualities are equally important for all visualization tasks.

Weight values may be any positive number (including 0.0). It is impossible to give suitable values for all Cxx for all possible cases *a priori*, but in general, most of the weight values can be set to 1.0 and only in some cases they should be increased or decreased according to the situation. The problem of finding suitable values for the weight values is similar to finding suitable values for the subqualities: they must be guessed after a careful analysis of the visualization context.

An example may be the layman visualization of daylight intensities in a proposed building. If a single lay person is to get an as comprehensive impression of the lighting situation as possible, an immersive inspection of the data using virtual reality technologies is very important and the weight value for presentation and interaction quality Cpi will be set to 1.0. If, on the other hand, a presentation to a large public via a magazine article is needed, the presentation and interaction quality is not important at all and Cpi has to be set to 0.0.

Thus, the visualization quality Q_{VIS} of a visualization according to a visualization context can be expressed by six pairs of two values each:

$$Q_{VIS} = [(C_{dr}, Q_{dr}); (C_{se}, Q_{se}); (C_{ma}, Q_{ma}); (C_{im}, Q_{im}); (C_{pi}, Q_{pi}); (C_{us}, Q_{us})] \tag{1}$$

These twelve numbers are *not* accumulated to one single number since this would mean a huge loss of information and since such an accumulated number would no longer allow one to achieve a fair comparison of different visualizations to each other.

In order to facilitate perception of the visualization quality as well as comparison of the qualities of two different visualizations, a visual representation of Q_{VIS} (called the 'Q_{VIS} graph') is used. It shows six vertical bars (one for each subquality). The height of each bar represents the subquality value Q, the width represents the weight value C. An example is shown in figure 2.

3 Example: Visualizing the Space Shuttle

The Q_{VIS} model and graph may be illustrated by looking at three different visualizations of NASA's space shuttle.

The following visualizations are compared:

1. a simple, static visualization created with the AVS[31, 1] system,
2. an interactive, distributed visualization done with the ISVAS system[8, 21], and
3. an immersive visualization using NASA AMES' Virtual Windtunnel (VWT)[5, 32].

Examples for the three visualizations (with user interfaces) are shown in figure 1. The image for VWT shows a plane instead of a shuttle data set, but visualizations of the space shuttle data have also been done in this system.

Comparing three visualizations done with these three systems is very difficult. The systems and visualizations differ quite a lot from each other; they were created having very diverse tasks (and thus visualization contexts) in mind.

First, the three visualization(systems) are briefly introduced. Afterwards, they are compared using Q_{VIS} graphs.

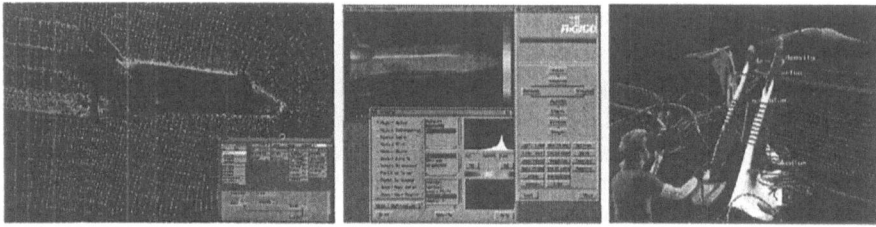

Fig. 1. Example pictures of the three shuttle visualizations with user interface (from left): static visualization with AVS, interactive distributed visualization with ISVAS, and immersive visualization (of a plane) with the Virtual Windtunnel (VWT) (picture with kind permission of NASA Ames).

3.1 The Three Visualization Systems

Static Visualization with AVS: AVS 5.0 is a very popular general purpose visualization system. It belongs th the class of application builders, i.e., users can configure their own visualization pipeline by arranging existing (or custom made) modules in a graphic interactive user interface.

The system can be use for interactive visualization. A PostScript output module also allows high quality static visualizations for printing.

For this paper, a static, printed visualization of space shuttle data is chosen as visualization context.

More information on AVS may be found in [31] and [1].

Interactive Distributed Visualization with ISVAS: ISVAS 3.2 is a flexible, monolithic visualization system which has been developed by Fraunhofer IGD since 1991. The main motivation for this work was the flexible visualization of large datasets in realtime.

The main focus of this software is on the visualization of FE (finite element) data, e.g., for structure dynamics or for fluid dynamics. System features which were introduced for this purpose include:

1. a very flexible calculator for complex operations on raw data, e.g., scaling of time varying tensor data or the combination of 3 scalar fields into one vector field;
2. interpolation in space (using shape functions) and time between given node values;
3. mapping functions of values to color, vector arrows, deformed geometry, etc.;
4. slicing, particle tracing, iso-surfaces in unstructured grids;
5. comparison of computed and measured values (e.g., strain on a steel shaft under load).

Another ISVAS data type is voxel data, e.g., medical MRI data, ultrasonic data, seismic data. Iso-surfaces and arbitrary slices are possible with this data type.

ISVAS can be coupled to simulation systems in order to allow online visualization. It also has been coupled successfully to a VR system, thus realizing immersive scientific visualization[15]. Furthermore, it allows collaborative, distributed visualization among two users.

More information on ISVAS may be found in [8] and [21].

Immersive visualization with the Virtual Windtunnel (VWT): In [5] and [32], a monolithic system for exploring numerically generated 3D unsteady flow fields is presented which employs virtual environment techniques. The system was designed for the very purpose of "walk around inside three-dimensional single grid steady flow tracking a streamline from the hand at frame rates" [6]. When it was presented in 1992 it was revolutionary in the way that it allows investigation of flow fields in VEs at reasonable frame rates. The fact that it has been developped by NASA Ames Research and the applications it is being used for (e.g., flow around Space Shuttle) made it clear that VE techniques indeed can be used for applicatins other than architectural walk throughts.

Yet, by trying to gain maximum performance, a very special system was designed which lacks many techniques used for scientific visualization or for virtual environments, e.g. level of detail.

3.2 Comparisons of the Three Visualizations

Now for a comparison of the three visualizations. Unfortunately, there is not enough space in this article to describe in detail the visualization contexts and the visualizations that led to the following Q_{VIS} graphs. Still, it is important to stress that the following is *not* an objective comparison of the three systems, but it is the comparison of three very specific visualizations done with these systems according to different demands (and visualization contexts).

The basis for the following comparisons were the visualization contexts that were shortly mentioned in the previous section. Both the weight values C as well as the subquality values Q had to be guessed by the author; extensive user surveys and testing would have led to more accurate results.

Table 1 gives the weight values and the subquality values if each visualization is rated according to its own visualization context. These values are visualized in the Q_{VIS} graph in fig. 2.

Table 1. Subquality values (Q) and weight values (C) for the three space shuttle visualizations, each according to its own visualization context. (Q_{VIS} graphs in fig. 2).

subquality		AVS		ISVAS		VWT	
		weight	quality	weight	quality	weight	quality
name	abbr.	C	Q	C	Q	C	Q
data resolution	dr	1	0.9	1	0.9	1	0.9
semantic quality	se	0	0	1	0.5	0.5	0
mapping quality	ma	1	0.9	1	0.9	1	0.8
image quality	im	1	1	1	0.8	0.8	0.8
presentat./interact.	pi	0.2	0.2	1	0.9	2	0.9
multi-user quality	us	0	0	1.5	1	1	0.8

It can be seen that for the static, printed visualization with AVS (leftmost graph), semantic and multi-user quality are completely unimportant, and presentation interaction quality also does not have significance (well, maybe only for generating the visualization it is preferred to have an interactive system instead of a batch oriented one, but not for consuming this visualization). The demands for image quality are fulfilled completely, those for data resolution and mapping quality almost, only presentation interaction quality is not too good.

Similarly, for distributed interactive visualization with ISVAS, multi-user quality is very important, but the other subqualities are also important. Data semantics is not met too well, since there is just static semantics in our test case, but online visualization with steering of the simulation process would be best. This is possible with ISVAS, but not realized for the space shuttle example. Multi-user quality meets the demands of the visualization context very well.

For the Virtual Windtunnel example, it must be admitted that the author did not have all the information that would be needed to make a very good Q_{VIS} evaluation. Some items had to be guessed. Semantic quality is not too important here, but it would be desirable to have online visualization in this example, which (to the knowledge of the author) is not the case. Of course, presentation interaction quality is very important and very good in this example of immersive scientific visualization.

So a look at the three Q_{VIS} graphs in figure 2 easily shows strengths and weaknesses as well as different demands of the three test cases.

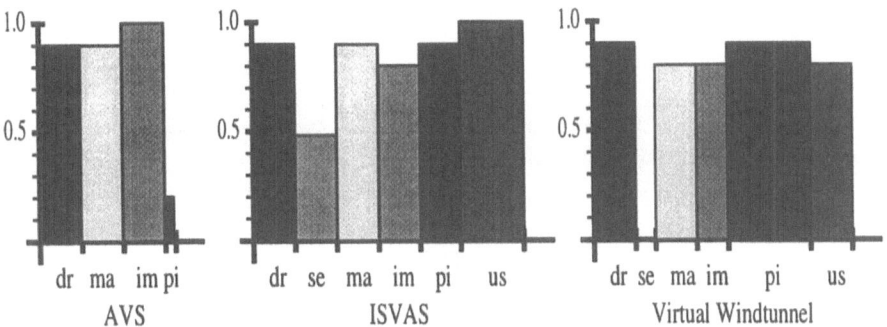

Fig. 2. Q_{VIS} graphs for the visualization quality of the three visualizations, each according to its own visualization context (see table 1).

Fig. 3, on the other hand, shows the Q_{VIS} graphs of the weight values and subquality values for the three visualizations if they are all evaluated according to a uniform, specific goal and visualization context: Two engineers want to discuss with each other on several complex vortexes in the flow field around the space shuttle.

Since the demands to the three visualizations are the same, the weight values Cxx also are equal for each of the three cases. Now, the static visualization with AVS does not meet the requirements very well. The requirements could be met much better with AVS if a different visualization would be done, but for the sake of this comparison, let's use the visualization that was done for the demands outlined for figure 2.

The graph for ISVAS has changed only a little, since the demands now are very similar to the ones of the previous example. The graph for VWT has changed more because the demands have changed. It can easily be seen in figure 3 that the described visualization with ISVAS best meets the demands that are now the same for each of the three visualizations, but this is only due to the fact that the demands are very similar to the ones that this ISVAS visualization had been designed for, and quite different from the initial demands of the two other visualizations.

4 Conclusion

The Q_{VIS} reference model is an approach to measure and to compare the quality of visualization systems or of visualizations by quantifying a number of sub-

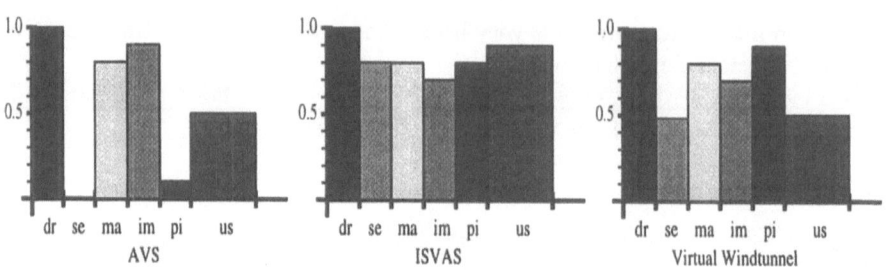

Fig. 3. Q_{VIS} graphs for the visualization quality of the three visualizations, all according to the same visualization context.

qualities as well as corresponding 'weight values', i.e., their importance. It is emphasized that the specific visualization context must be considered.

This visualization context includes the prior knowledge of the user, her or his visualization aims, the kind of application, the amount and structure of the data to visualize, as well as the available hard- and software.

The mentioned subqualities are data resolution, semantic quality, mapping quality, image quality, presentation and interaction quality, and multi-user quality, which sometimes are an assembly of several subitems. The Q_{VIS} graph is a compact, easy to perceive and to compare representation of the so-defined visualization quality.

Please note that the definition of the quality of a visualization *system* in section 2.1 does not take into account the flexibility or robustness of a visualization system to behave nice under a variety of different user demands (e.g., a variety of users with different demands). So a future extension of the reference model woud be to evaluate these aspects of a visualization system.

The reference model can help to evaluate the quality of a visualization or of a visualization system. Still, it must be stressed that the 'perfect' visualization (system) does not exist and cannot exist. The individual demands of the users, but also the changing aims of a single user and the data sets to visualize are too heterogenious. Furthermore, some requirements for an optimal visualization are contrary to each other and will never be harmonized completely. An example for such contrasting demands is the wish to achieve high frame rates in visualizing exponentially growing data sets in sometimes very high quality representation, if possible even over large distances. No matter how the performance of hard- and software should evolve in the future, it is clear that such demands always will require compromises.

Still, the proposed reference model for visualization quality does not specify a completely objective metrics – to achieve this would be an irrational goal considering the very individual properties of visualization quality – but it is a first approximation of a good tool for evaluating and comparing visualizations and visualization systems. This can eventually lead to improved visualizations and thus to more or faster insights into raw data and underlying phenomena.

Computer scientists will have to work together with users (e.g., engineers) of visualizations and they will have to learn from teachers, advertisement experts, designers, and artists who have investigated the best usage of visualization (color, shape, and many more aspects) for a long time. The findings of all of these people must not be ignored if we further want to improve the quality of our scientific visualizations.

Acknowledgements

The author wishes to thank the paper referees for some valuable comments. Many thanks also to Prof. J.L. Encarnação, Stefan Müller, and Florian Schröder, who provided the environment which made this work possible.

References

1. http://www.avs.com/products/avs5/index.html
2. Baker, P.: Knowledge-based visualization, *IEEE Visualization '92 Workshop on Automated Design of Visualizations*, Boston, MA, October 1992
3. Baker, M.P., Bushell, C.: After the Storm: Consiferations for Information Visualization, *IEEE Computer Graphics and Applications*, May 1995, pp. 12–15
4. Beshers, C., Feiner, S.: Automated Design of Data Visualizations, in: Rosenblum, L., Earnshaw, R., et al. (eds.): *Scientific Visualization – Advances and Challenges*, Academic Press, London, 1994, pp. 87–102
5. Bryson, S., Levit, C.: *The Virtual Windtunnel*, IEEE Computer Graphics and Applications, 12, 4, pp. 25-34, 1992
6. Bryson, S., Levit, C.: *Lessons learned while implementing the virtual windtunnel project*, Visualization '92, Tutorial # 2, 4.1–4.7, 1992
7. Foley, J.D., van Dam, A., Feiner, S.K., Hughes, J.F.: *Computer graphics: principles and practice*, Addison-Wesley, Reading, Mass., 1990
8. Frühauf, T., Göbel, M., Haase, H., Karlsson, K.: Design of a Flexible Monolithic Visualization System, in: Rosenblum, L., Earnshaw, R., et al. (eds.): *Scientific Visualization – Advances and Challenges*, Academic Press, London, 1994, pp. 265–286
9. Gerfelder, N., Müller, M.: Quality Aspects of Computer-Based Video Services, in: SMPTE, *Proceedings of 1994 European SMPTE Conference*, Cologne, September 1994, pp. 44–67
10. Gershon, N.: From Perception to Visualization, in: Rosenblum, L., Earnshaw, R., et al. (eds.): *Scientific Visualization – Advances and Challenges*, Academic Press, London, 1994, pp. 129–139
11. Globus, A., Raible, E.: Fourteen Ways to Say Nothing with Scientific Visualization, *IEEE Computer*, Vol. 27, No. 7, July 1994, pp. 86–66
12. Globus, A., Uselton, S.: Evaluation of Visualization Software, *Computer Graphics*, May 1995, pp. 41–44
13. Haase, H., Dohrmann, C.: Doing It Right: Psychological Tests to Ensure the Quality of Scientific Visualization, in: Göbel, M., David, J., Slavik, P., van Wijk, J.J. (eds): *Virtual Environements and Scientific Visualization '96*, Springer Verlag, 1996, pp. 243-256

14. Haase, H., Press, T.: Improved Interaction and Visualization of Finite Element Data for Virtual Prototyping, *Proc. ASME International Computers in Engineering Conference*, Sacramento, USA, September 1997

15. Haase, H.: Symbiosis of Virtual Reality and Scientific Visualization System, *Computer Graphics Forum*, Vol. 15, No. 3, August 1996

16. Haase, H.: Evaluating the Quality of Scientific Visualizations: The Q-VIS Reference Model, in: *Proc. SPIE Visual Data Exploartion and Analysis V Conference*, paper no. 3298-18, San Jose, CA, January 1998

17. Haase, H., Strassner, J., Dai, F.: Virtual Molecules, Rendering Speed, and Image Quality, in: Göbel, M. (eds.): *Virtual Environments '95*, Springer Verlag, Wien, 1995, pp. 70-86 and pp. 296-298

18. Haase, H., Strassner, J., Dai, F.: VR Techniques for the Investigation of Molecule Data, *Computers & Graphics*, Elsevier Science Ltd., Vol. 20, Nr. 2, 1996, pp. 207-217

19. Haase, H., Dai, F., Strassner, J., Göbel, M.: Immersive Investigation of Scientific Data, in: Nielson, G., et al. (eds): *Scientific Visualization: Overviews, Methodologies & Techniques*, IEEE Computer Society Press, 1997

20. Hibbard, W., Dyer, C., Paul, B.: Display of scientific data structures for algorithm visualization, *Proceedings IEEE Visualization '92*, IEEE Computer Society Press, October 1992, pp. 139-146

21. http://www.igd.fhg.de/www/igd-a4/projects/docs/isvas/

22. Jung, V.: Fuzzy Effectiveness Evaluation for Intelligent User Interfaces to GIS Visualization, *Proceedings Fourth ACM Workshop on Advances in Geographic Information Systems*, Rockville, Maryland, Nov. 1996, ACM Press, New York, 1996, pp. 157-164

23. Keller, P.R., Keller, M.M.: *Visual Cues – Practical Data Visualization*, IEEE Computer Society Press, Los Alamitos, 1993

24. Mackinley, J: Automatic Design of Graphical Presentations of Relational Information, *ACM Transactions on Graphics*, Vol. 5, Nr. 2, April 1986, pp. 110-141

25. Miceli, K., Domik, G.: A visualization framework for multidisciplinary data analysis, *IEEE Visualization '92 Workshop on Automated Design of Visualizations*, Boston, MA, October 1992

26. Robertson, P.K., De Ferrari, L.: Systematic approaches to visualization : is a reference model needed? in: Rosenblum, L., Earnshaw, R., et al. (eds.): *Scientific Visualization – Advances and Challenges*, Academic Press, London, 1994, pp. 87-102

27. Robertson, P.K.: A methodology for scientific data visualization: Choosing representations based on a natural scene paradigm, *Proceedings IEEE Visualization '90*, IEEE Computer Society Press, October 1990, pp. 114-123

28. Rogowitz, B., Treinish, L.: Data structures and perceptual structures, in: Rogowitz, B., Allebach, J. (eds.): *SPIE Proceedings: Human Vision, Visual Processing and Digital Display IV*, SPIE, Vol. 1913, 1993

29. Robertson, P.K., Silver, D.: Visualization Case Studies: Completing the Loop, IEEE Computer Graphics and Applications, May 1995, pp. 18-19

30. Senay, H., Ignatius, E.: Compositional analysis and synthesis of scientific data visualization techniques, in: Patrikalakis, N.M. (eds.): *Scientific Visualization of Physical Phenomena (Proceedings CG International '91)*, Springer-Verlag, Tokyo, pp. 269-281

31. VandeWettering, M.: The Application Visualization System – AVS 2.0, in: *Pixel*, July/August 1990, pp. 30-33

32. http://www-sci.nas.nasa.gov:80/Software/VWT/

Editor's Note: see Appendix, p. 150 for colored figures of this paper

Three-Dimensional Visualization of Atomic Collision Cascades

Filip Šroubek and Pavel Slavík

Department of Computer Science and Engineering
Czech Technical University Prague, Czech Republic

Abstract. The paper describes a new approach to the visualization of atomic collision cascades and using the interaction with visualized data. The collision cascade is a physical phenomenon initiated by bombarding the surface of a solid with accelerated atomic particles. The process evolves in time and therefore it is necessary to develop some tools that would allow to investigate and visualize the dynamics of the process. Such tools are classifiers (filters) that enable to select and visualize objects with specific dynamic properties. As the visualization has been done in a 3D environment a question arises how to specify effectively and user friendly both the properties and the objects in the 3D space. Several techniques are available that allow interaction in the 3D space. It has been necessary to test some of these techniques and to determine which one is the most suitable for the given application class.

1   Introduction

A simulation of real physical systems is usually a time consuming problem connected with a tedious visualization task. Many such physical systems are described by a set of time dependent differential equations. Using computers we are able to solve these equation sets and afterwards we may visualize the results so that scientists can get a better insight into the studied physical processes. The problem we have been solving is a non-linear molecular-dynamics system with a large number of first-order differential equations. Dynamic systems are in general described by $\frac{d\boldsymbol{X}}{dt} = \boldsymbol{F}(\boldsymbol{X}, \boldsymbol{\alpha}, t)$ equations where $\boldsymbol{X} \in \mathbb{R}^n$ of the n-dimensional phase space, $\boldsymbol{\alpha}$ are system constant parameters and t is the time. By solving these equations we obtain $\boldsymbol{X}(t)$. The equations can be only solved by means of numerical methods. The dimension n of the phase space as of $\boldsymbol{X}(t)$ is usually very high and a projection into a less-dimensional space is inevitable, e.g. visualization of molecular-dynamics would be a projection into a 4D space (3D Euler's space and time parameter). In Sect. 2 we discuss the molecular-dynamics and its computer simulation in more detail. Section 3 describes our approach of molecular-dynamics visualization with two-level classifiers that help to reduce the number of visualized data. Special attention has been paid to the handling of dynamic attributes of the visualized objects. Section 4 describes our experiment that enables to compare the effectiveness of the interaction in different display modes. In the implementation we have restricted ourselves to common visualization devices; anaglyph glasses, shutter stereo glasses as view devices and a common mouse device as an 2D input device. We did not work with more advanced and more expensive VR tools.

2 Physical Background of the Problem

The physical phenomenon in our case could be shortly described as follows. The surface of a solid e.g. aluminum is bombarded with accelerated ions or neutrals. During the collision the accelerated particle interacts with atoms of the solid and knocks some of them out of the surface. This dynamic process is called the *collision cascade*. The scientists are especially interested in the way how the surface atoms are knocked out (the process is called the atomic *sputtering*), in which direction they are emitted and what kinetic energy they have. The information on sputtered particles is of a significant importance in several surface analytical techniques, in techniques used for cleaning of solid surfaces and in modern technologies based on ion induced chemical processes (e.g. mechanical hardening of polymer surfaces). Also of the interest is the average distance which the bombarded particles travel in the solids. The knowledge of this range is of great practical interest in the technological process called implantation in which impurity atoms are implanted in semiconductors to change their resistivity in a controlled way or into metals to change their mechanical properties.

Dynamic properties of particles are in general described by a second-order differential equation of the following form

$$\boldsymbol{F} = m\frac{d^2\boldsymbol{r}}{dt^2} = -\operatorname{grad}\varphi\,, \tag{1}$$

where $\boldsymbol{F}$ is the force acting on the particle, m is the particle mass, $\boldsymbol{r}$ is the position vector and φ is the potential at the point $\boldsymbol{r}$. The potential φ describes how the particles interact with each other. The precision of the simulation depends critically upon a proper choice of this potential. The simplest is the *binary potential* which depends only upon the particles between which it is calculated and on the distance between them. It is independent of positions of the other particles. The potential function is usually a composition of more than one function joined together, each of which is valid only within a specific distance interval. The value of the binary potential is significant only within the certain radius and outside this radius it is assumed to be zero (a finite cut-off radius). Some examples of binary potential could be find in [1] and [2]. The (1) describes the dynamic properties of one particle and could be split into 6 simple first-order equations of the type $\frac{dv_i}{dt} = -\frac{1}{m_i}\frac{d\varphi_i}{dr}$ and $\frac{dr_i}{dt} = v_i$, where v_i is the velocity and r_i is the position vector of the i-th particle. In our case we investigate the collision cascades produced by relatively low energy primary particles (< 1 keV). It turns out that for such low impact energies the collision cascades occupy a small volume and thus clusters typically 10x10x10 atoms are sufficient. The number of equations to be solved is then around 6000. The dynamic process is over in about 200 fs after the impact of the primary particles. For the purpose of the simulation we have implemented a one-step numerical method for solving first-order equations called *Runge-Kutta* with Gill's modifications [5] [6] (we call the program **SPUTT**).

2.1 Statistics

In actual experiments, when the surface of solids is bombarded with accelerated particles, the surface is not hit exactly at the same place several times. Instead, we have a stream of accelerated particles, which collide with a large area of the surface. On the other hand in the software simulation of collisions (in the **SPUTT** program) the particle hits the surface precisely at the point given by the user. For a single crystal with a perfect surface at temperature zero, the surface crystallographic structure defines an *irreducible surface element*. Using the symmetry and translation invariance of the surface, the ion bombardment into this irreducible surface element is representative of the ion bombardment of the entire surface. So we solve the collision cascades for different impacts on a regular grid in the irreducible surface element. We have denoted the obtained collision cascades as a *collision cascade set* (**CCS**).

2.2 Simulation Time Requirements

To give an example of the computing time of common problems the simulation of a 10x10x8 Al cluster bombarded with 560 eV Ar ion at 1000 different impact points took 48 hours on IBM RS6000 workstation. There are several techniques how to reduce the computation time of the simulation. Firstly, the number of atoms with which any particle interacts is finite due to the fact that we have a finite cut-off radius of the potential. Since the identity of the interaction partners changes with time as a consequence of particle motion, neighbor lists are used. The best list nowadays in use is the Verlet-linked-cells algorithm [7]. Secondly, it is preferred to encode potentials as look-up tables rather than as analytical expressions to be evaluated. Thirdly, the numerical method for solving differential equations can take advantage of determining a local error in each time step and appropriately increasing or decreasing the time step. Implementing all three techniques we may reduce the computation time by a factor of 10.

3 Visualization of Molecular-Dynamics

Visualization is an essential part of many dynamic system simulations. The molecular-dynamics simulation of sputtering as it was described above is not an exception. Information about the sputtered yield are of a great importance and most of the scientific monographs concerning the molecular-dynamics provide a statistical analysis of simulated data and then compare obtained results with laboratory experiments. Some of the common analyses are: the distributions of the sputtered yield within the irreducible surface element (see Sect. 2.1) [3], the energy and polar-angular distributions of sputtered particles [1], etc. Direct visualization of the bombardment is often presented in these monographs in the form of static images of collision cascades before and/or after the bombardment, see [3] and [4].

We have aimed our research at a direct dynamic visualization of collision cascades *CASVIS* (*CAS*cade *VIS*ualization), i.e. we track and display individual particles in time during collisions. Dynamic visualization of CCS as a whole would inevitably involve special statistical approaches that would project collision cascades in CCS into some more comprehensive data. One of the approaches could be to display "temperature" distribution in CCS as a function of time and thus be able to track the "*hot spots*" (areas of high activity). Here the "temperature" $\mathcal{T}(r, t)$ at a certain point r and time t could be defined as a mean value of energies of particles averaged over all collision cascades in CCS located at time t in an area around point r. This approach is in a stage of development and we want to cover it in our future work. Iconic techniques [10] that display particular parameters at given points of a more complex visualization procedure may be of a great benefit here, e.g. by displaying sputter yields at specified points.

Up to now, we visualize collision cascades in CCS separately and we have not implemented any statistical procedures. Unfortunately this technique may obscure some features otherwise observable in statistical approaches, but on the other hand it gives a precise insight into dynamic properties of individual collision cascades. We know that some impacts trigger dramatic sputter activities and such processes are only observable by visualizing suspected collision cascades. Knowledge of atomic-dynamics during coincident collision cascades, when sputter and scatter phenomena occur simultaneously, is of a great importance for scientists. Such cascades are selected from CCS and properly visualized only by means of classifiers.

3.1 Classifiers

As we have seen in Sect. 2.2 one CCS consists of 1000 collision cascades each for one cluster of around 1000 particles. We face a common problem in the scientific visualization of physical processes, i.e. we are overwhelmed by a vast number of data. Use of adjustable software classifiers that assort collision cascades in CCS is thus convenient. Two level classification is proposed in this paper. Classifiers of the first level, i.e. the *cascade classifiers*, are designed to extract collision cascades from CCS that match some given specification. Classifiers of the second level, i.e. the *particle classifiers*, are intended to select (emphasize) "interesting" particles in the visualized collision cascade. By the word "interesting" we mean particles that match certain conditions.

Applying cascade classifiers on CCS produces a *collision cascade subset*. An example of two cascade classifiers follows:

- **Select collision cascades with a certain sputter yield.**
  We can, for example, select cascades with a high or a low sputter activity, or select cascades in which only certain particles of the cluster are sputtered.

– **Select collision cascades in which impact particles penetrate to a certain depth of the cluster.**

  We can, for example, determine cascades in which an accelerated impact particle bounces off the cluster surface, i.e. the impact particle is *scattered* by the cluster.

An efficiency of the cluster classifier can be designed as $1 - (N_{sel}/N_{all})$, where N_{all} is the number of all collision cascades in CCS, and N_{sel} is the number of cascades in CCS selected by the classifier. The efficiency depends completely on the type of the cluster classifier and on the type of CCS. Currently we have implemented one cluster classifier that selects clusters in which an impact particle bounces off the surface. Applying this classifier on CCS containing 450 collision of a 945 eV Li ion with an Al cluster has led to 255 matches.

The particle classifiers are included in the user-interface of our visualization system CASVIS. So far we have implemented two particle classifiers in the following way:

– **Select particles with energies above a certain threshold**

  This way we select particles which will move faster than the specified threshold of the kinetic energy. CASVIS features an easy setting of the threshold with a scrollbar widget. In Fig. 2 we see the result of this classifier. Particles that match the classification are drawn as spheres and the rest as dots.

– **Select particles that will pass through a certain plane**

  If we set the plane above and parallel to the cluster surface we select particles that will be sputtered through that plane. CASVIS enables easy manipulation of the plane by moving and rotating the qplane object. In Fig. 3 we see the selected particles drawn as spheres and the plane object.

In CASVIS users can interactively set parameters of the classifiers and thus dynamically modify the visualization output according to the users' needs, i.e. the given approach allows us to describe interactively some dynamic properties of the visualized objects. These properties form a special class of classifiers (filters) that help to visualize the dynamics of the process in certain context and thus lead to better understanding of the visualized physical phenomenon. It is obvious that the setting of classifier parameters requires an intensive interaction in 3D space due to the 3D nature of classifier characteristics.

3.2 Features of CASVIS

In our case of the many-particle problem we do not have to solve issues connected with visualization of vector or scalar fields, although the potential field created by particles is of a scientific interest and in our future work we want to implement it. We perceive individual particles as points in space that behave according to well defined rules, see Sect. 2 and therefore we have decided to represent them simply as spheres. CASVIS implements three different 3D display modes: *perspective viewing*, *anaglyph stereo* and *shutter glass stereo*, as follows:

- **perspective viewing**

  A perspective projection transformation is applied to the image data (a position of particles). Particles are projected onto a 2D plane in color. No additional hardware is required.

- **anaglyph stereo**

  Two perspective views of clusters are generated; for right and left eye with complementary colors (red/blue or red/green). 3D perception is achieved through anaglyph glasses with filters in complementary colors. The serious disadvantage of this method is the lack of color information. This information is only partially present in shades of grey.

- **shutter glass stereo**

  Two perspective views of clusters are generated using the same color scale as in the perspective viewing. The right and left eye views are presented alternatively on the screen.

CASVIS behaves identically under all three display modes. We restricted ourselves to 2D input device mouse and use a so called *laser beam* [11] to assist interaction in the 3D scene. The beam is a ray, cast from the user's pointer in a straight line perpendicular to the screen plane. The first object to be intersected by the ray is selected for manipulation. This way users can drag and rotate the particle cluster in space and select particles. In this manner selected particles display their current position and energy. Due to a large number of displayed particles, we have decided to plot selected particles as spheres and remaining ones as dots. This technique dramatically improves the perception because significant areas are emphasized. To improve and speed up the selection a selecting technique of *particle classifiers* introduced in Sect. 3.1 was implemented, e.g. pick up particles with energies above a specified threshold (Fig. 2) or pick up particles that were sputtered (Fig. 3). Colors are used to distinguish between different kinetic energy levels of particles. Kinetic energies are mapped into a color spectrum and the mapping can be adjusted arbitrarily. Finally trajectories of selected particles can be also included in the image. Users can track the cascade development by changing the time manually or by turning on an animation sequence. In Fig. 4 we have shown the collision cascade 140 fs after the impact.

4 Psychological Test of the Interaction Effectiveness in Different Display Modes

Up to now we were mainly concerned with the question how to map simulated data to meaningful images that would help to understand collision cascades. One of the interesting issues connected with it is the effectiveness of a particular interaction technique in a specific display mode. This issue is in general very broad and complex. Many experiments were proposed in the past and carried out in different environments and with emphasizes on different tasks.

Unfortunately, up to now, only a static perception of visualized 3D objects was performed. In [8] images of chemical molecules were used and volunteers (subjects) were asked to perform three different tasks (identifying, comparison, movement) under three different display modes (perspective, anaglyph stereo, shutter glass stereo). Two interesting conclusions were reached. Firstly, if the third dimension doesn't provide any necessary information the interpretation of the visualization is more difficult. Secondly, the viewing in the anaglyph mode is as good as in the shutter mode. Presented with a pair of object images, the subject's task in [9] was to determine whether the two images represented the same or different objects. One of the interesting results was that the accuracy of the subject's task was significantly improved when the object motion was controlled. However the response time of the subject's task was longer.

4.1 Experiment

The above described experiments were of passive nature. In our study subjects, besides the image interpretation, performed also the *interaction* in 3D. The purpose was to evaluate the effectiveness of three display modes for different interaction tasks. These interaction tasks were selected as typical for the given application. CASVIS system discussed in Sect. 3 served only as an experimental environment. The scientific interpretation of images was not considered in our experiment. The experiment involved 20 subjects who were students of Computer Science and had more or less the same experience with the 3D representation. The tasks were performed interactively on each of the 3D display modes (perspective, anaglyph stereo, shutter glass stereo). Presented with a cluster of particles (spheres) positioned in 10x10x8 cube, the subject's tasks were the following:

1. set the "plane" object parallel to the larger cube sides,
2. selecting individual particles define 2x2x2 cube in one corner of the 10x10x8 cube,
3. count selected particles (spheres) in the cube.

As an interaction tool, the mouse with a *laser beam* selection technique was used, see Sect. 3.2 for a detail description. Tasks were designed to take the advantage of the classifiers and to emphasize different domains; task 1 should emphasize the object manipulation in space, task 2 the object selection in space and task 3 the identification of objects. In all the cases the subjects were allowed to rotate the cluster in order to improve the accuracy [9]. We were aware of the fact that the subjects could improve their performance during testing because they can find (learn) more efficient procedures how to reach the desired output of the tests. To reduce the influence of this learning parameter we allowed the subjects to perform the tasks more than once before measuring their response time. The accuracy of a subject's response was measured for task 1 and 3 only.

4.2 Results and Discussion

The results of the experiment should show in which display modes the 3D interaction is the most suitable to solve the particular task. Table 1 summarizes

Table 1. Mean response time $\bar{t}$ and mean response error $\bar{e}$ of 20 subjects for three different tasks and three display modes. F and p-value were computed by a one-way Analysis of Variance (ANOVA) and are defined in [12]

	mean response	Persp.	Anaglyph	Shutter	$F(2,57)$	p-value
Task 1	$\bar{t}(s)$	105.87	116.87	74.25	4.36831	0.01636
	$\bar{e}(degrees)$	3.84	2.45	1.88	2.72328	0.07272
Task 2	$\bar{t}(s)$	27.37	40.12	26.75	2.09214	0.13118
Task 3	$\bar{t}(s/sphere)$	1.16	1.17	0.94	1.00434	0.37158
	$\bar{e}(\text{mismatched counts})$	0.5	0.37	1.75	5.28571	0.00867

the main effects of display modes obtained from our psychological test. Mean response times and mean response errors are plotted in Fig. 1 for individual tasks. F and p-value were computed by a one-way Analysis of Variance (ANOVA) for $\alpha = 0.05$. A detailed discussion of the statistical quantities can be found in [12] and will not be covered in our paper.

In task 1 the display mode has a significant influence on mean response times and mean response errors; p-value < 0.1 was found in both cases. The object manipulation in the shutter mode was performed more accurate and faster than in the perspective and the anaglyph modes, see Figs. 1(a), 1(b)). According to a Scheffé test [13] only the mean response time in the shutter mode was significantly better than in the anaglyph mode. This was an expected result.

Mean response times of task 2 and task 3 are not significantly influenced by the display mode (p-value > 0.1). So the observation (see Table 1 and Fig. 1(c)) that the shutter and the perspective modes are faster than the anaglyph mode for object selection and identifying is only a trend and not a significant result. On contrary counting of spheres in task 3 was significantly more accurate in the anaglyph and the perspective modes than in the shutter mode as the Scheffé test [13] indicated (see Fig. 1(d)). Results obtained from task 3 correlate with the results in [8], where the authors conclude that display modes have no significant influence on mean response times for counting rings in simple molecules. In our task 3 images consist of around 35 spheres and that was probably of similar complexity as images of simple molecules in [8]. Moreover, from task 2 we deduce that the same conclusion holds also for the object selection.

It is obvious, from the obtained results, that the 3D interaction is the most effective in the shutter display mode, but the 3D perception could be blurred in this mode. Furthermore the 3D interactive control of scientific visualization environment improves both understanding and manipulation with visualized objects. Therefore the investigation of new 3D interactive controls in scientific visualization may introduce desired positive effects. We believe that the particle classifiers introduced in Sect. 3.1 are the good example of such interactive controls.

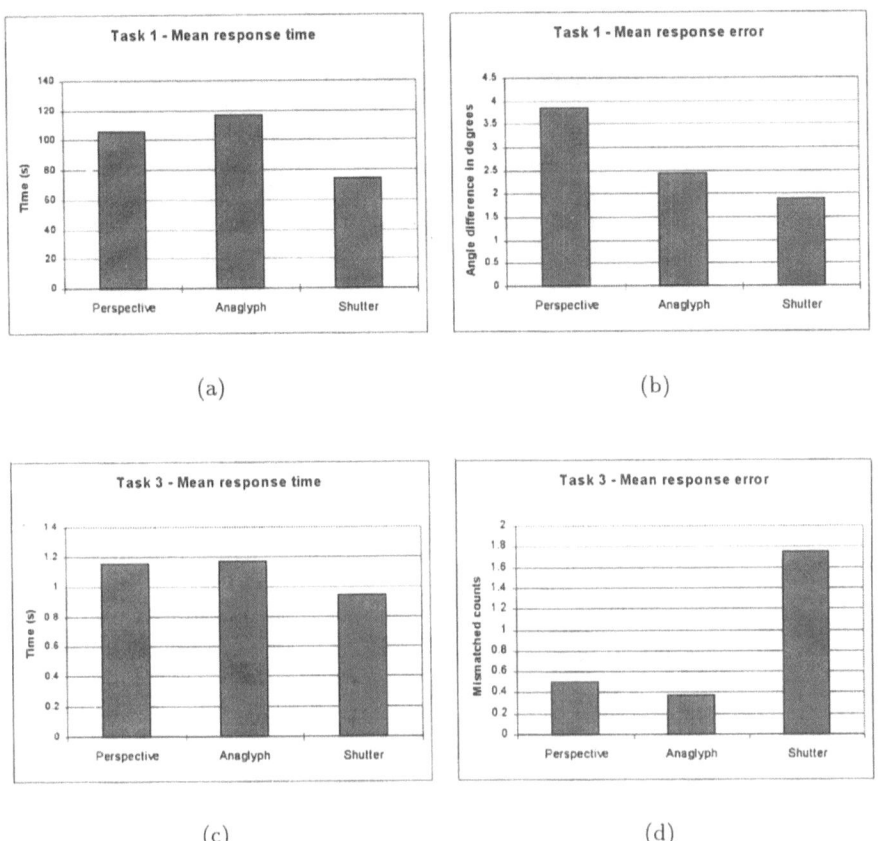

Fig. 1. Mean response times (in seconds) for task 1 - 1(a) , mean response errors (deviation of plane normal from correct orientation in degrees) for task 1 - 1(b); mean response times (in seconds/one sphere) for task 3 - 1(c), mean response errors (number of mismatched counts) for task 3 - 1(d)

References

1. M.H. Shapiro and J. Fine, Phys. Rev, B44 (1989) 43-53.
2. J.W. Hartman, M.H. Shapiro, T.A. Tombrello and J.A. Yarmoff, Phys. Rev. B.(February 1997).
3. H.M. Urbassek, Molecular-dynamics simulation of sputtering, Nucl. Instr. and Meth. in Phys. Res. B 122 (1997) 427-441.
4. D.N. Bernardo, R. Bhatia and B.J. Garrison, keV particle bombardment of solids: molecular dynamics simulation and beyond, Comput. Phys. Commun. 80 (1994) 259-273.
5. T.E. Shoup, Numerical Methods fot the Personal Computers, Prentice-Hall, Inc., 1983.

6. J.R. Rice, Numerical Methods, Software, and Analysis, McGraw-Hill, Inc., 1987.

7. G.S.Grest, B. Dünweg and K. Kremer, Comput. Phys. Commun. 55 (1989) 269.

8. S. Volbracht, K. Shahrbabaki, G. Domik and G. Fels, Perspective viewing, Anaglyph stereo or Shutter glass stereo?, Proc. of the IEEE Conference Visualization (96) 192-193.

9. G.S. Hubona, G.W. Shirah and D.G. Fout, 3D Object Recognition with Motion, Proc. CHI 97 ACM Press

10. R.-T. Happe and M. Rumpf, Characterizing Global Features of Simulation Data by Selected Local Icons, Proc. 7th Eurographics WorkShop on Visualization in Scientific Computing (1996) 80-88.

11. Chris Hand, A Survey of 3D Interaction Techniques, Computer Graphics forum Volume 16 (1997) 269-281.

12. One-way ANOVA,
    http://www.richland.cc.il.us/james/lecture/m170/ch13-1wy.html.

13. Scheffé test,
    http://www.richland.cc.il.us/james/lecture/m170/ch13-dif.html.

Editor's Note: see Appendix, p. 151 for colored figures of this paper

Color Plates

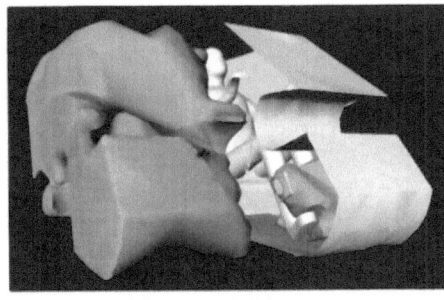 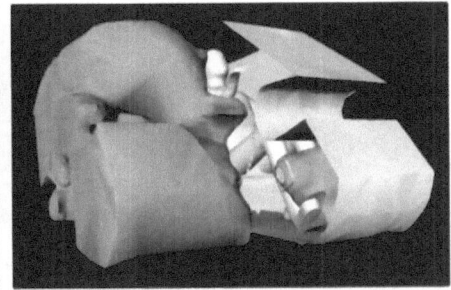

Isosurface reduced by the Vertex removal (left) and Edge collapse (right) strategies to 25% (12540 triangles) (Frank et.al., Fig. 5).

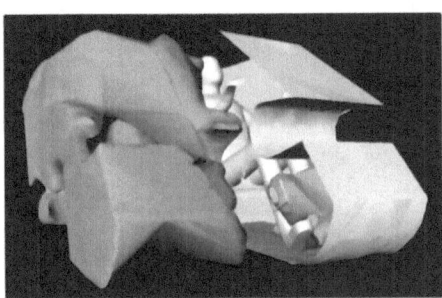 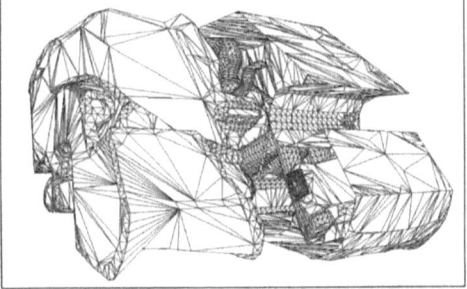

Isosurface reduced by the Vertex removal strategy to 20% (10032 triangles) (Frank et.al., Fig. 6).

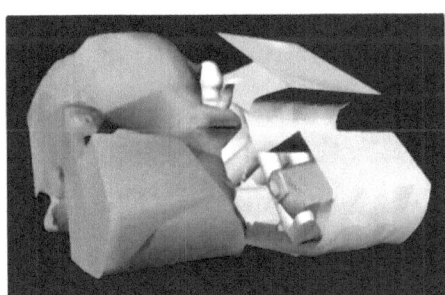 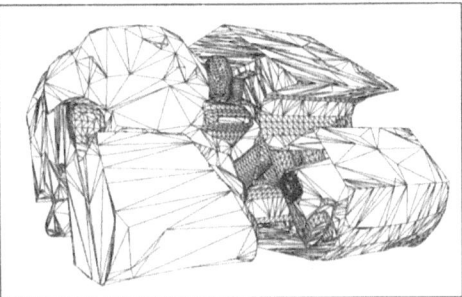

Isosurface reduced by the Edge collapse strategy to 15% (7523 triangles) (Frank et.al., Fig. 7).

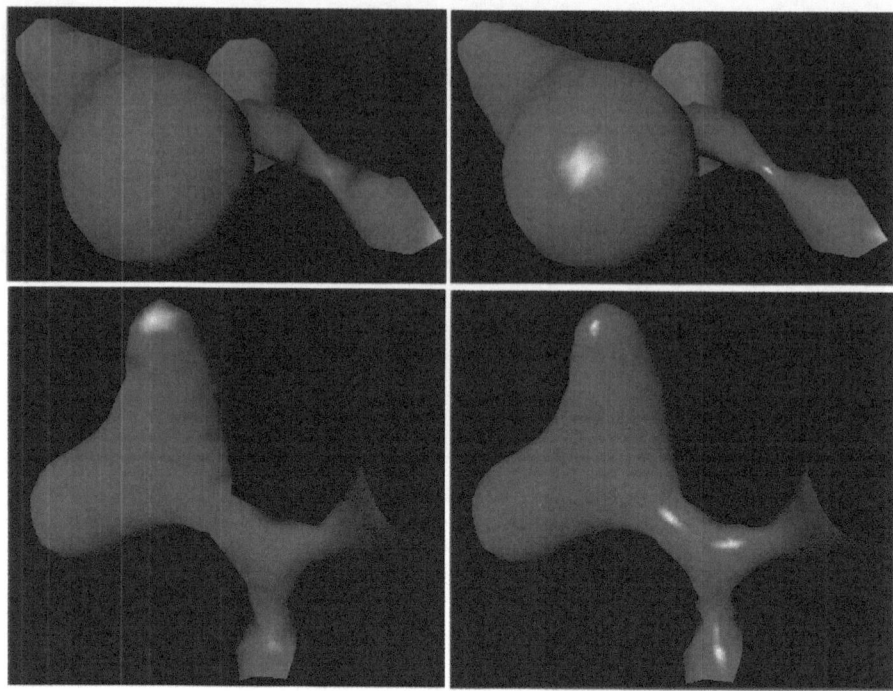

Two meshes extracted from a section of the SOD dataset, using standard MC on the left and the enhanced precision method on the right (max recursion level = 4); above are two top views and a side view is used for the images below (Allamandri et.al., Fig. 7).

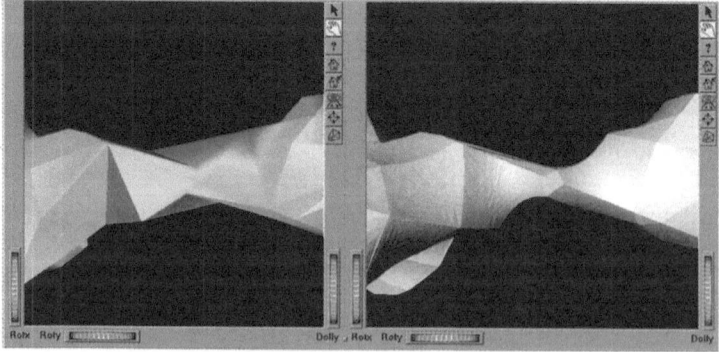

Comparing iso-surfaces extracted (SOD dataset) with the Metro tool; a section of the MC mesh (left image) is colored according to the distance from the corresponding mesh section extracted with PreciseMC (right image) (Allamandri et.al., Fig. 8).

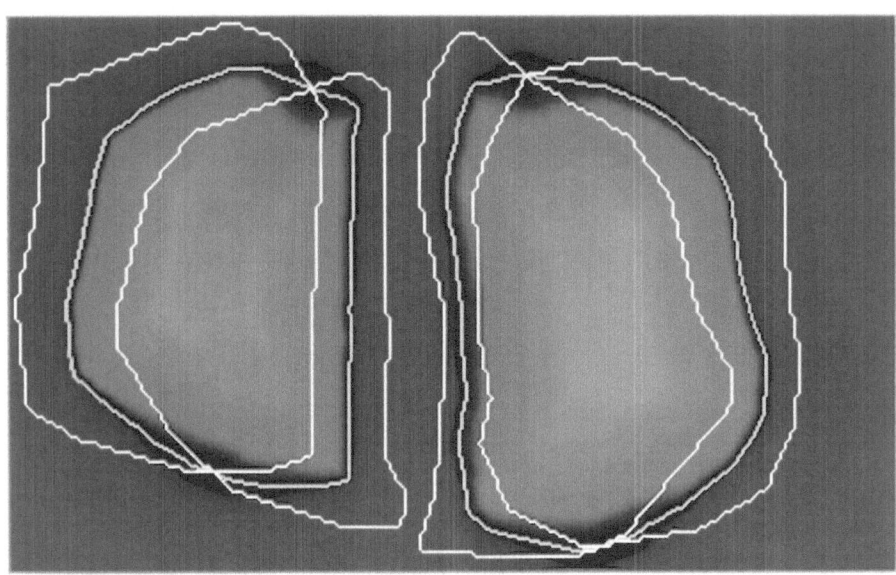

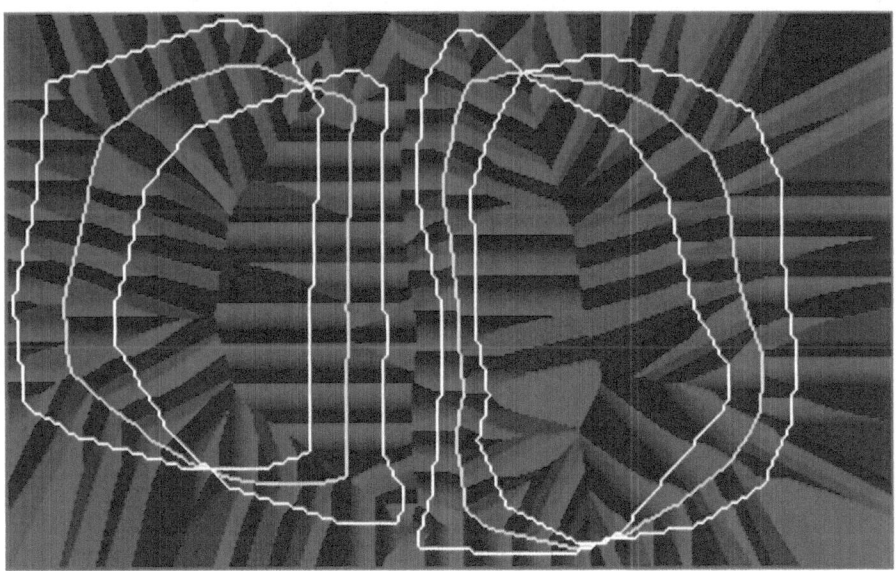

Sum of two distance fields of consecutive slices. Negative distance values are shown in green, positive distance values in blue. The border constitutes the medial axis (shown in yellow). The correspondences to the closest contourpixel are shown in red. (Schilling et.al., Fig. 10)

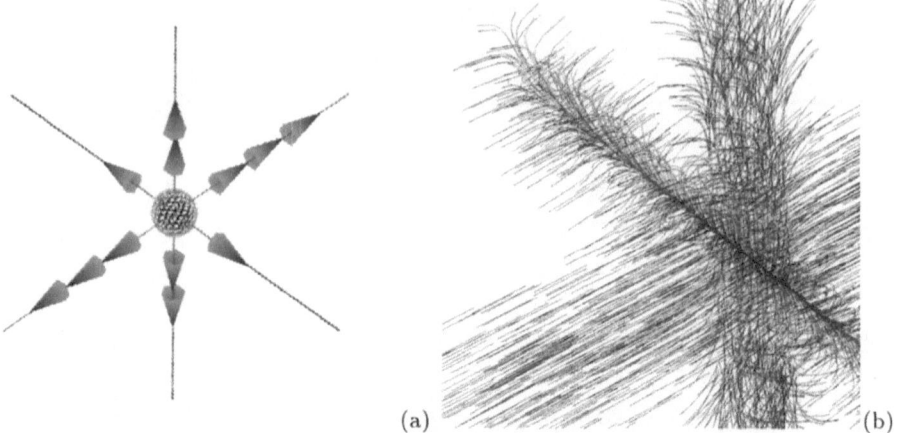

(a) (b)

Visualizing the flow near a linear node repellor in 3D: eigenvectors and eigenvalues (1, 10, and 100) (a), characteristic trajectories plus threads of streamlets (b) (Löffelmann et.al., Fig. 3).

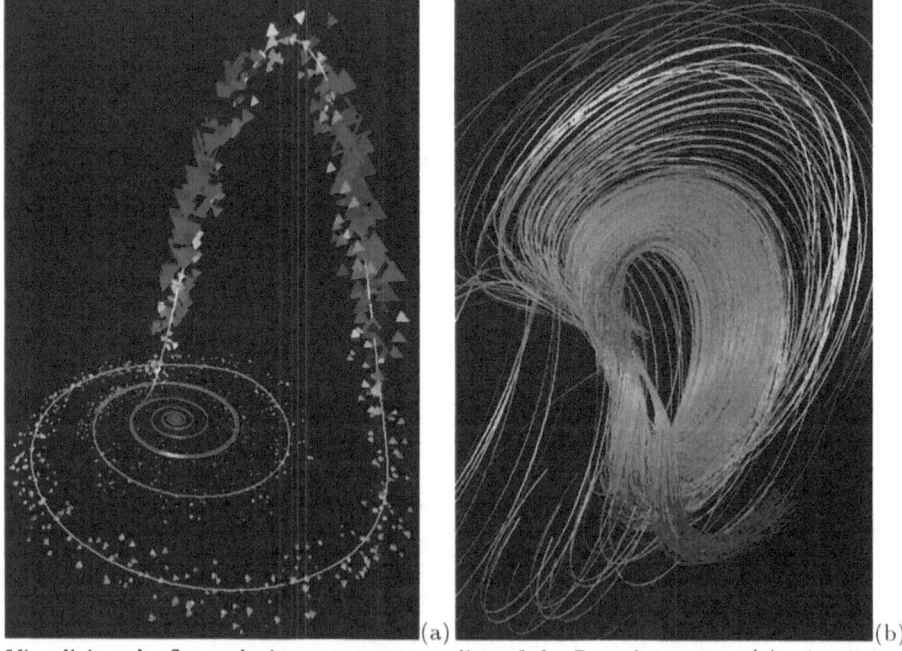

(a) (b)

Visualizing the flow velocity near a stream line of the Roessler system (a); visualizing the dynamics of a periodic dynamical system exhibiting a twisted torus (b) (Löffelmann et.al., Fig. 4).

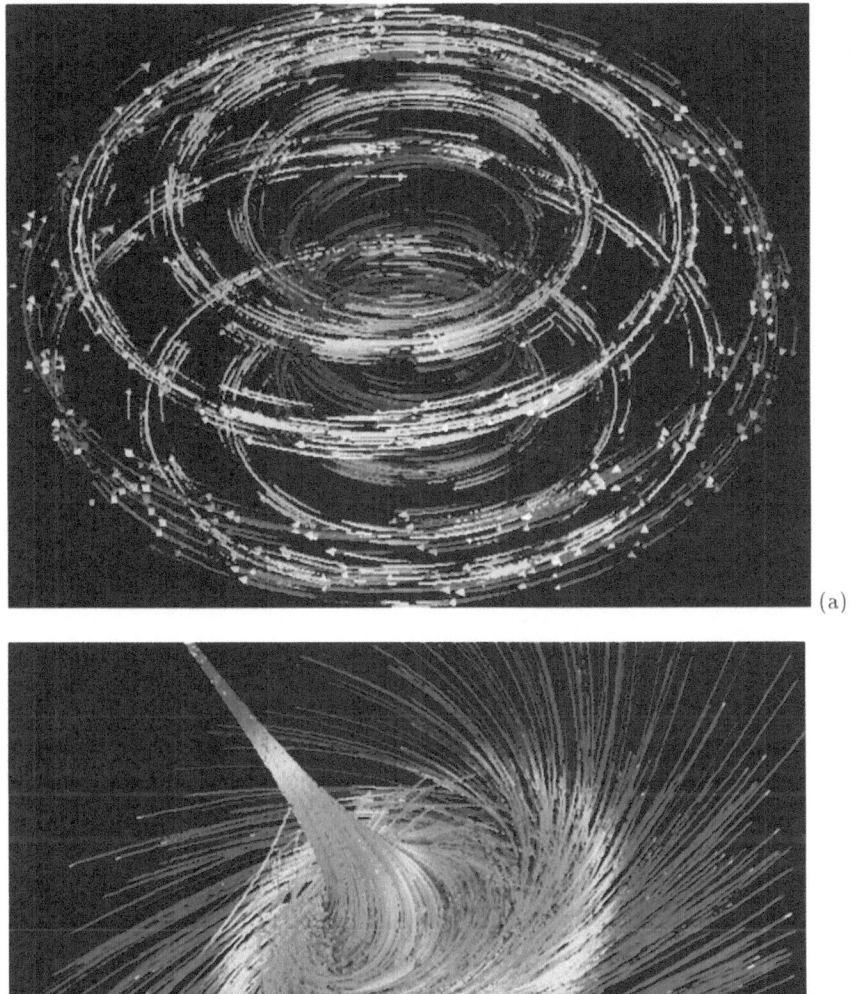

(a)

(b)

A thread of streamlets visualizing the flow near a torus in 3D space (a); flow near a 3D focus visualized using two threads of streamlets (b) (Löffelmann et.al., Fig. 5).

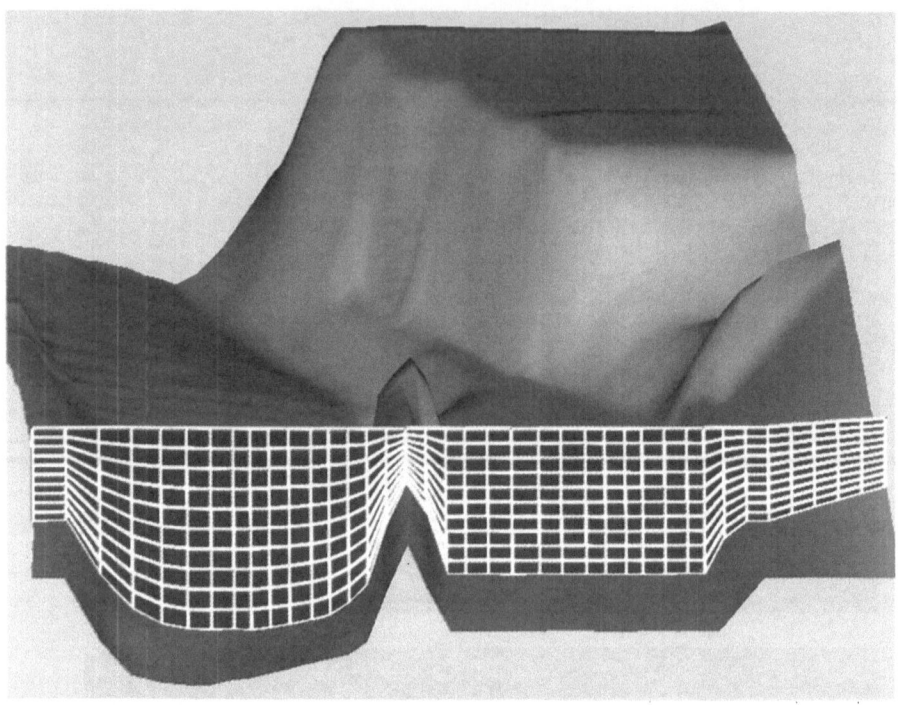

Sigma-transformed grid in Lith harbour (Sadarjoen et.al., Fig. 9).

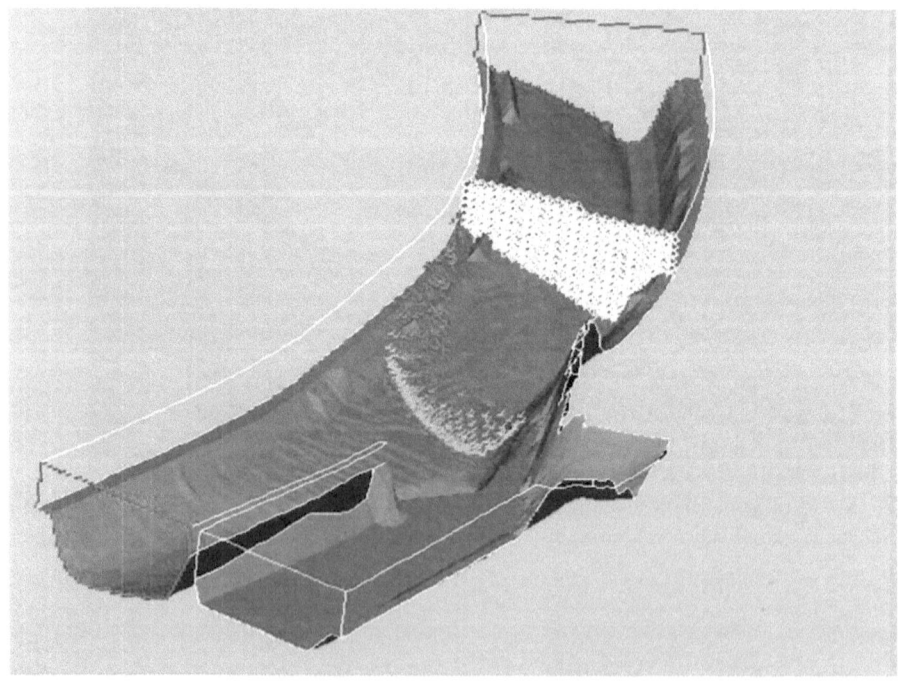

Particles successfully traced with the 6-decomposition method
(Sadarjoen et.al., Fig. 10).

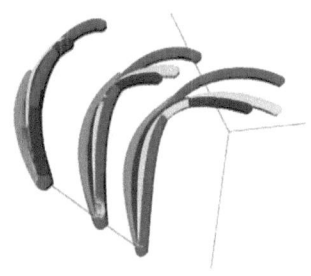

Colored streak balls and tetrahedra in a vortex flow given on a sparse grid (Teitzel et al., Fig. 3)

Streak tubes in a cavity flow; the red tubes are computed on a full grid of level 7, the other tubes are created on sparse grids of level 7 (yellow), 5 (blue), and 3 (green) (Teitzel et al., Fig. 4)

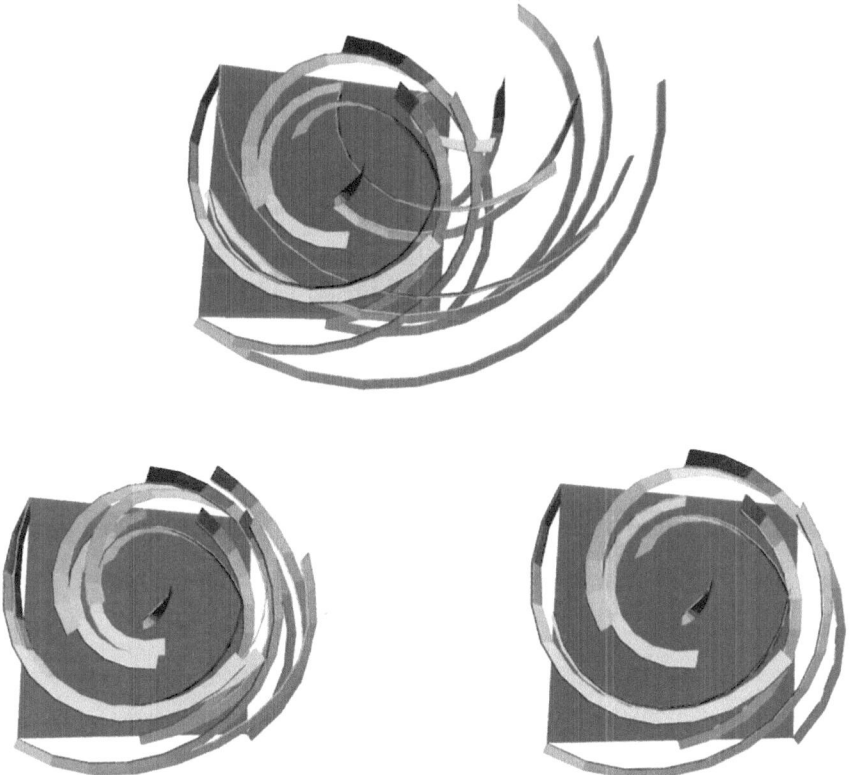

Streak bands in a vortex flow; ribbons containing blue edges display the flow on a full grid of level 7, bands with green edges the flow on sparse grids of level 0 (top), 1 (left), and 4 (right); the ribbons computed on full and sparse grids coincide on screen for levels greater than 3 (Teitzel et. al., Fig. 5)

148

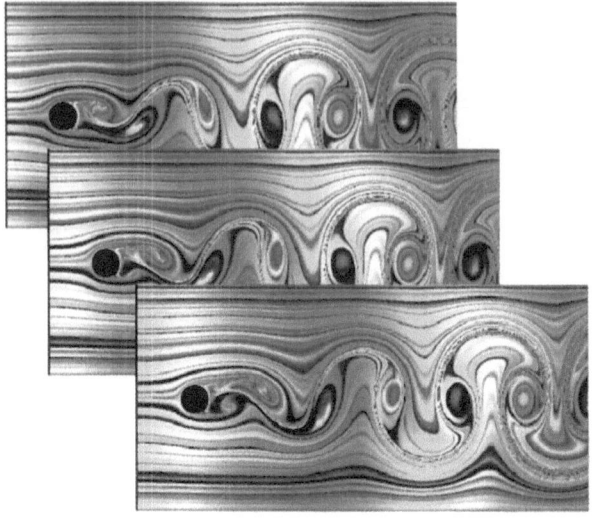

Texture transport in the von Kármán vortex street (Becker et.al., Fig. 6).

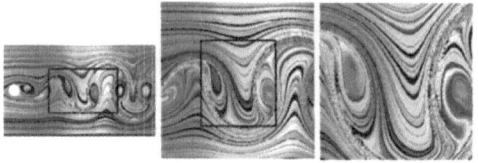

Several intermediate steps in a continuous zoom into the physical space Ω
(Becker et.al., Fig. 7).

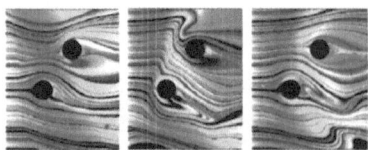

Texture transport applied to a compressible flow arround two cylinders
(Becker et.al., Fig. 8).

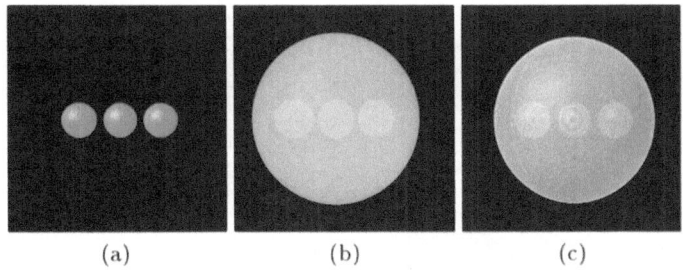

(a) Inner-only, (b) transparent, (c) surface (Hubbold et.al., Fig. 4).

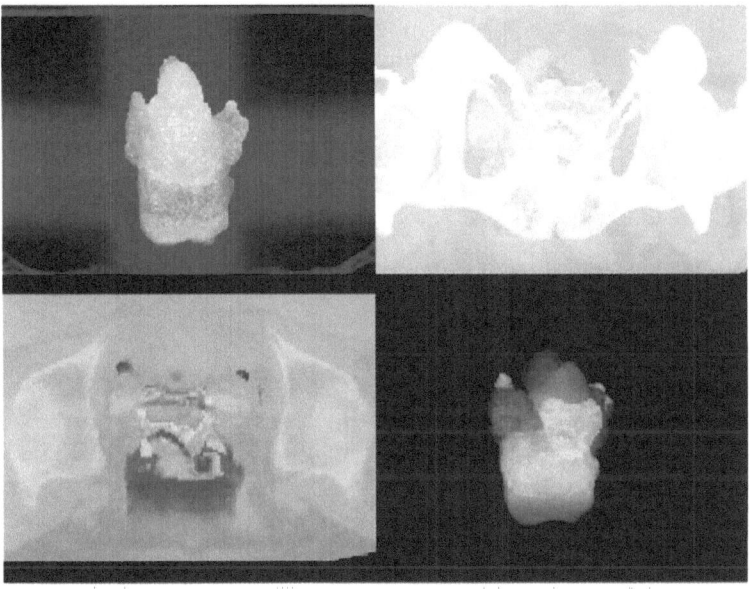

Four views composited into a single image. The compositing is performed in 3D, although this is not apparent in this monoscopic view (Hubbold et.al., Fig. 5).

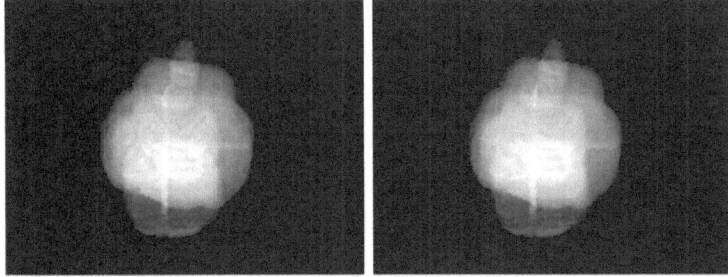

Stereoscopic pair showing nested features, arranged for cross-eyed viewing (Hubbold et.al., Fig. 6).

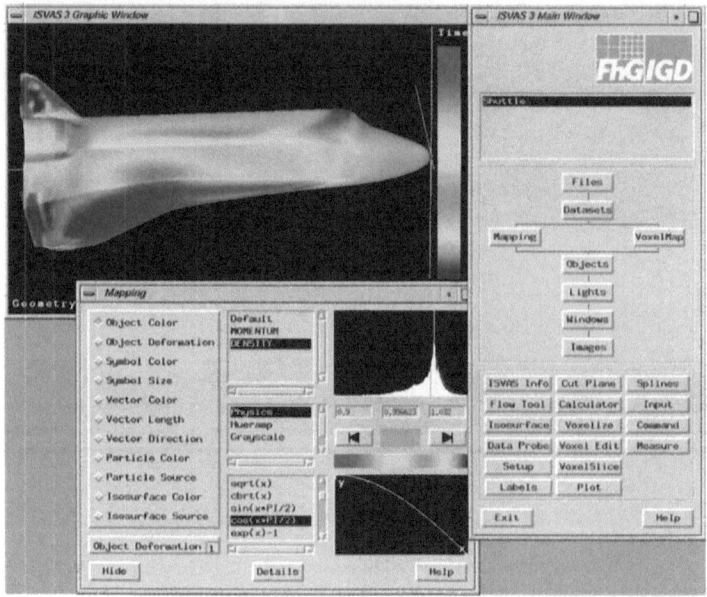

Screenshot of interactive distributed visualization with ISVAS (Haase, Fig. 4).

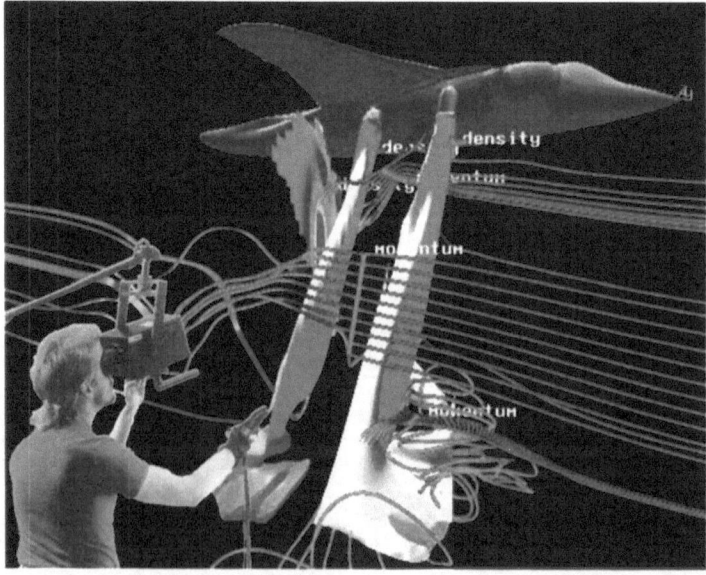

Example picture of immersive visualization with the Virtual Windtunnel (VWT) (picture with kind permission of NASA Ames) (Haase, Fig. 5).

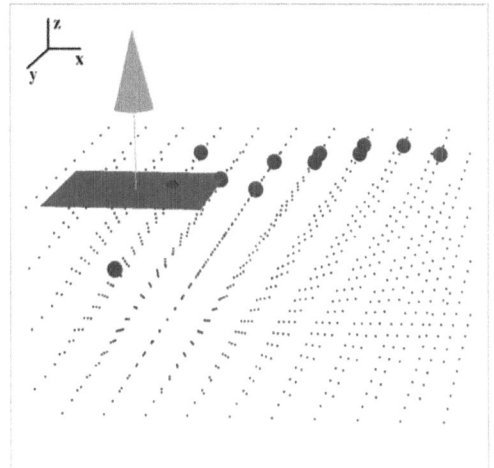

The Al cluster before the impact of the Ar ion (555 eV) as in Fig. 2. Particles, displayed as spheres, will sputter later in the collision cascade and are selected by the *plane particle classifier*. The plane object is the rectangle with the normal vector. The image was produced by CASVIS in the perspective display mode (Šroubek et.al., Fig.3).

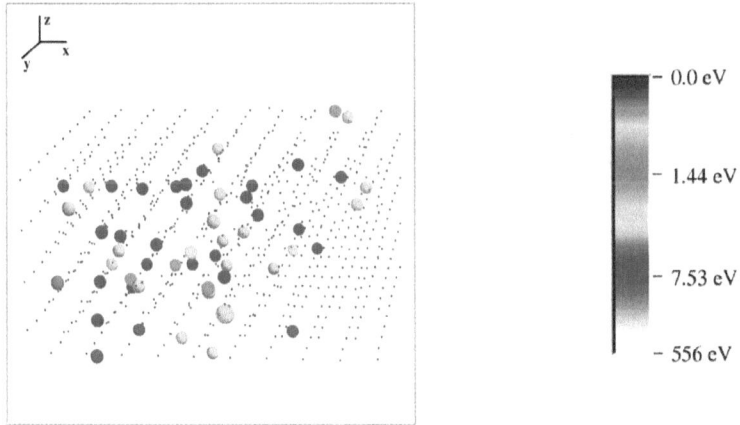

The Al cluster 140 fs after the impact of the Ar ion (555 eV). The image was produced by CASVIS in the perspective display mode. The *energy particle classifier* is set to 5 eV. Color mapping of kinetic energies is according to the color bar on the right (Šroubek et.al., Fig.4).

SpringerEurographics

George Drettakis,

Nelson Max (eds.)

Rendering Techniques '98

Proceedings of the Eurographics
Workshop in Vienna, Austria,
June 29–July 1, 1998

1998. 231 partly coloured figures. XI, 339 pages.
Soft cover DM 118,–, öS 826,–
ISBN 3-211-83213-0. Eurographics

Some of the best current research on realistic
rendering is included in this volume. It
emphasizes the current "hot topics" in this
field: image based rendering, and efficient
local and global-illumination calculations. In
the first of these areas, there are several contributions on real-world model acquisition
and display, on using image-based techniques for illumination and on efficient ways
to parameterize and compress images or light
fields, as well as on clever uses of texture and
compositing hardware to achieve image
warping and 3D surface textures. In global
and local illumination, there are contributions on extending the techniques beyond
diffuse reflections, to include specular and
more general angle dependent reflection
functions, on efficiently representing and
approximating these reflection functions, on
representing light sources and on approximating visibility and shadows. Finally, there
are two contributions on how to use knowledge about human perception to concentrate
the work of accurate rendering only where it
will be noticed, and a survey of computer
graphics techniques used in the production
of a feature length computer-animated film
with full 3D characters.

Panos Markopoulos,

Peter Johnson (eds.)

Design, Specification
and Verification
of Interactive Systems '98

Proceedings of the Eurographics
Workshop in Abingdon, U.K.,
June 3–5, 1998

1998. Numerous figures. Approx. 310 pages.
Soft cover approx. DM 118,–, öS 826,–
ISBN 3-211-83212-2. Eurographics

Does modelling, formal or otherwise, have a
role to play in designing interactive systems?
A proliferation of interactive devices and
technologies are used in an ever increasing
diversity of contexts and combinations in
professional and every-day life. This
development poses a significant challenge
to modelling approaches used for the design
of interactive systems. The papers in this
volume discuss a range of modelling
approaches, the representations they use,
the strengths and weaknesses of their
associated specification and analysis techniques and their role in supporting the design
of interactive systems.

SpringerWienNewYork

Sachsenplatz 4-6, P.O.Box 89, A-1201 Wien, Fax +43-1-330 24 26
e-mail: order@springer.at. Internet: http://www.springer.at
New York, NY 10010, 175 Fifth Avenue • D-14197 Berlin, Heidelberger Platz 3
Tokyo 113, 3-13, Hongo 3-chome, Bunkyo-ku